APPAREL MANUFACTURING MANAGEMENT SYSTEMS

TEXTILE SERIES

Howard L. Needles, Editor

FABRIC FORMING SYSTEMS
By *Peter Schwartz, Trever Rhodes* and *Mansour Mohamed*

TEXTILE IDENTIFICATION, CONSERVATION, AND PRESERVATION
By *Rosalie Rosso King*

TEXTILE MARKETING MANAGEMENT
By *Gordon A. Berkstresser III*

TEXTILE WET PROCESSES: Preparation of Fibers and Fabrics
By *Edward S. Olson*

TEXTILE FIBERS, DYES, FINISHES, AND PROCESSES: A Concise Guide
By *Howard L. Needles*

AUTOMATION AND ROBOTICS IN THE TEXTILE AND APPAREL INDUSTRIES
Edited by *Gordon A. Berkstresser III* and *David R. Buchanan*

APPAREL MANUFACTURING SYSTEMS: A Computer-Oriented Approach
By *Edwin M. McPherson*

Other Title

TEXTILE DYEING OPERATIONS: Chemistry, Equipment, Procedures, and Environmental Aspects
By *S.V. Kulkarni, C.D. Blackwell, A.L. Blackard, C.W. Stackhouse* and *M.W. Alexander*

APPAREL MANUFACTURING MANAGEMENT SYSTEMS

A Computer-Oriented Approach

by

Edwin M. McPherson

School of Textiles
North Carolina State University
Raleigh, North Carolina

np

NOYES PUBLICATIONS
Park Ridge, New Jersey, U.S.A.

Library of Congress Catalog Card Number: 87-12239
ISBN: 0-8155-1141-8
Printed in the United States

Published in the United States of America by
Noyes Publications
Mill Road, Park Ridge, New Jersey 07656

10 9 8 7 6 5 4 3 2 1

Library of Congress Cataloging-in-Publication Data

McPherson, Edwin M.
Apparel manufacturing management systems.

Includes index.
1. Clothing trade--Management--Data processing.
I. Title.
HD9940.A2M42 1987 687'.068 87-12239
ISBN 0-8155-1141-8

Preface

The body of systems and control knowledge has been in a state of rapid growth with the advent of faster and lower cost computers. Nowhere has this been more apparent than in the apparel industry. To now, there has not been an effort to analyze and define this systems activity although there have been articles and seminars on various aspects of the emerging control systems. This book has used those seminars, lectures, articles, and specialized management systems publications of the apparel industry as a basis to merge the pieces into a unified view of all this activity.

The emerging information age is in the process of redefining how businesses will be run as well as restructuring the organization and control of the apparel industry, and the computer is multiplying the effectiveness of management when it is placed into proper use. In this book, we look at the past control systems and then at the emerging information age. Information flow is tracked in a novel way to provide a tool for rapid analysis of data transmission and its use. Basic apparel manufacturing system controls are detailed. The impact of linked computers is discussed. A system for reducing plant manufacturing time lags to improve customer service and reduce inventories is also outlined. Also discussed is the use of engineering data developed from payroll sources. Apparel plant managers are given a practical blueprint as to how they can control inventory and operating costs. Financial, merchandising and manufacturing control systems are placed into relationship with each other.

The text of this book was prepared at the request of the Management Systems Committee of the American Apparel Manufacturers Association. It was felt that it should have three purposes: (1) text book use; (2) a middle management handbook on systems; and (3) required reading reference for new AAMA management information personnel.

It is my hope that this book will meet these purposes.

Raleigh, North Carolina
April, 1987

Edwin M. McPherson

NOTICE

Contents and Subject Index

1

Introduction

INFORMATION TECHNOLOGY

Information technology consists of a convergence of computers and communications. This blending of electronic systems represents this country's strongest competitive advantage to offset the influx of low cost labor products from lesser developed countries. Advances in communications have changed the financial, economic and political functions of the whole world. Information technology will change everything that has ever been done in operating a business.

In the style-oriented, customer-driven apparel business, information technology changes the way the fiber, textile, apparel and retail businesses relate to each other. Opportunities for new service concepts abound. Vertical product integration can now be achieved through appropriate use of information technology rather than only through merger.

Multiple layers of management may no longer be necessary since information technology permits the gathering of information from any location at any time and the dissemination of that information at the same time in an appropriate format to anyone who needs it for instant decision making. Just as customer activity data can be collected by scanning retail sales, so too can that data be reflected at every level of the Fiber, Textile and Apparel Complex.

Information technology changes an entire organizational structure. This technology permits bosses at every level to review production, sales and financial data. Properly applied, information flow can be processed automatically into useful decision-making formats. The use of these data flows becomes embedded within a corporation. It invents a whole new thinking process.

THE INFORMATION AGE

Information handling has replaced manufacturing as the most important economic activity in the United States. Retailers, apparel, textile and fiber producers are increasingly linking into each other's systems. Under the American Apparel Manufacturers Association (AAMA) sponsorship, a Textile Apparel Linkage Council has been formed. This effort represents one of the several linkage efforts between industry segments. Many major retailers are linking with their suppliers. Involved in establishing such linkages is the prosaic definition of terms, formats and language upon which the manufacturing electronic infrastructure must be based.

Telecommunications has come to be recognized as the essential infrastructure for the emerging information age. User needs in information processing are becoming even more pervasive and varied. This electronic infrastructure has changed or is changing the world in a way that impacts not only the business world but also the economic and political environment. The power of television camera in conjunction with telecommunications reversed the recent election of Marcos to the Philippine presidency. This did not occur in previous Philippine elections not covered live.

The same force presents a never-ending variety of living conditions, people and apparel styles to the world. It can be anticipated that changes in expectations of people will be encouraged by the electronic age. This provides new opportunities to industry. However, the ability of people to manage information (receive, store, process and transmit) is still limited. The computer and modern telecommunications (phones, wired systems, cable, satellites) are actively helping people to overcome many information management problems. The more information people have, the more they use it innovatively. Companies that empower their staff with information technology will be strengthened.

Information systems are an important source of competitive advantage. Proper internal control of information together with the increasing use of externally available information is now a necessity. Internal communication networks have reduced or eliminated the barrier of distance in relaying operating information for decision making.

INFORMATION EXPECTATIONS

Only a few years ago there were repeated articles in the Business Press on Local Area Networks (LAN). The implications of these articles was that buildings, campuses, factories, etc., should be cable-wired so that any employee could plug in a terminal and manipulate data. When the concept was first proposed, it was not available. Today, with the inclusion of fiber optics and other transmission capabilities, the technology is in wide use.

Today, expectations for information services include such things as shopping by TV, automatic alarm services, instant convenient access to internal and external data bases, etc. Under the private enterprise system, the technology and entrepreneurs are ready to meet these needs. Information services of all types are growing daily. These services range from financial through legal data, from libraries to cable networks offering items for sale. Expectations are in the process of being satisfied.

Businesses and individuals are able to choose the technologies that meet their needs and their budgets. The use of these technologies will increase efficiency and productivity. Many who wish to work at home and "telecommute" are now using communication technology to do so.

Expert systems (or artificial intelligence) designed to put the knowledge of a human expert into a computerized format have generated widespread attention, just as LAN did a few years ago. These systems are in the process of moving from academia into the business world on micro computers. To some extent "expert systems" is considered a "buzzword" among computer professionals. Like many buzzwords, the meaning is ill-defined, but the expectation that the expert system will be useful is high. Expert systems will succeed in commercial markets when they become compatible with present EDP systems. They can then be integrated into existing software and systems. Their niche appears to be the incorporation of knowledge systems as enhancements to programs and systems which already exist. They may do such things as loan evaluations for financial services or equipment evaluations for manufacturers. Many artificial intelligence systems are being developed in the diagnosis area.

Expert systems can be viewed as offering new ways of preserving perishable human expertise, of distributing expertise, or fusing multiple sources of knowledge, of converting knowledge into a competitive edge and of altering business perspectives.

MANAGING KNOWLEDGE

Managing knowledge increasingly means managing electronically delivered information. The screening, selecting and manipulation of data is one of the most important tasks facing corporate management today. The advent of the personal computer (PC) has made timely information available to corporate executives that their predecessors did not have.

Top corporate officers can now track by computer a variety of factors that may affect their decisions, such as:

- Run a check on a company that is buying their stock.
- Check on a firm they may be considering for purchase.
- Follow political events in countries where they have investments.
- Keep up with latest trends in prices of raw materials.
- Check on current plans of firms for expansion or contraction of capacity.

Not only are data services such as Dunns, stock market data, various wire services (UP, AP, NEXIS, etc.) available, but personal computers can be linked to corporate main frames so that internal data may be reviewed. The curtain has risen on a whole new opportunity for managing information by computer.

JUDGING A COMPANY

Many firms judge their executives by their "bottom line" performance. Data on sales, collections, debts, inventories, etc., are all summarized in the annual report. Reasons for low profits, sales changes, and other activity are often stated in the annual report text. Modern data processing systems are geared to produce "bottom line" data by division, by plant, by sales unit, for the company as a whole so that responsible executives can act on this data. One simple analysis follows:

	Albany Intern'l (000)	Bendix (000,000)
Report Year	1982	1982
Annual Sales	$332,687	$4,092
Daily Sales	911.5	11.2
Days of Accts. Receiv.	58.4	51
Days of Inventory	91	68

	Kellwood (000)	Example of Good Performance
Report Year	1985	
Annual Sales	$588,531	-
Daily Sales	1,612	-
Days of Accts. Receiv.	41	45
Days of Inventory	88	60

Another Measure (000)

Debt	÷	Equity	
$123,750		$148,893	Debt to Equity Ratio .83 to 1

These two examples typify accounting interpretation of data. This type of data accumulation is now being fused into the new information technology.

At one time computers were primarily used by the finance department (or perhaps a research and development group). With the advent of small, relatively inexpensive computers, the use of computers has proliferated. They are used for everything ranging from word processing through controlling sewing equipment. To understand the basis of the information age, one must study its roots and the evolution of computer programs and systems into current systems. From these investigations one can then extrapolate how businesses may be affected in the future.

INDUSTRY CHARACTERISTICS

The use of computer systems within industry as a whole and within apparel in particular is in a state of flux. However, a generalized picture can be drawn to represent some relationships in respect to the apparel activity position in industry in general.

Figure 1.1, Industry Spectrum, illustrates the relation between types of industry and investment. In general, the larger the plant investment, the larger the expenditure upon interconnected control devices. There is constant movement within most areas of industry from left to right on the chart.

Handicraft industry, when profitable, attracts investment, engineering talent and gradually moves into an industry based upon more and more complex equipment. Semi-automated industry, under pressures of competition, evolves raw materals, equipment and volumes warranting further automation. The lack of one or more elements which justify larger equipment investment results in industrial stagnation, low margins and paper thin profits. Figure 1.2, Equipment versus Type

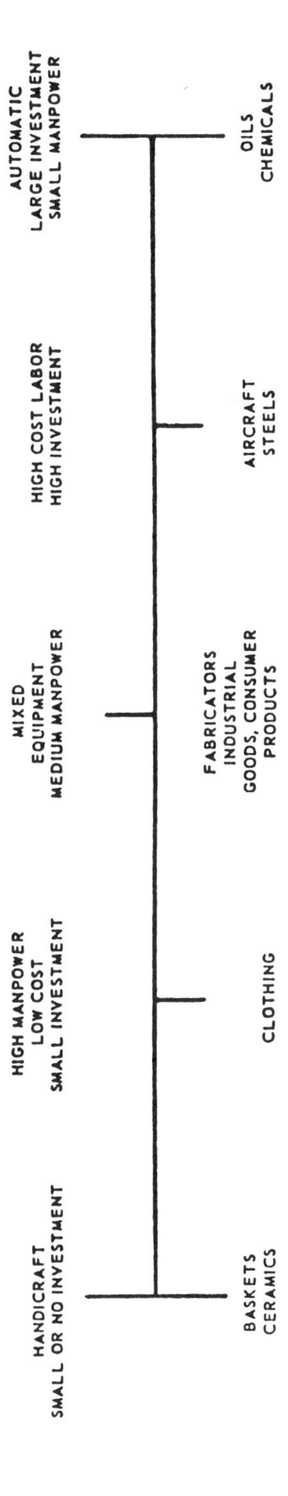

FIGURE 1.1. INDUSTRY SPECTRUM

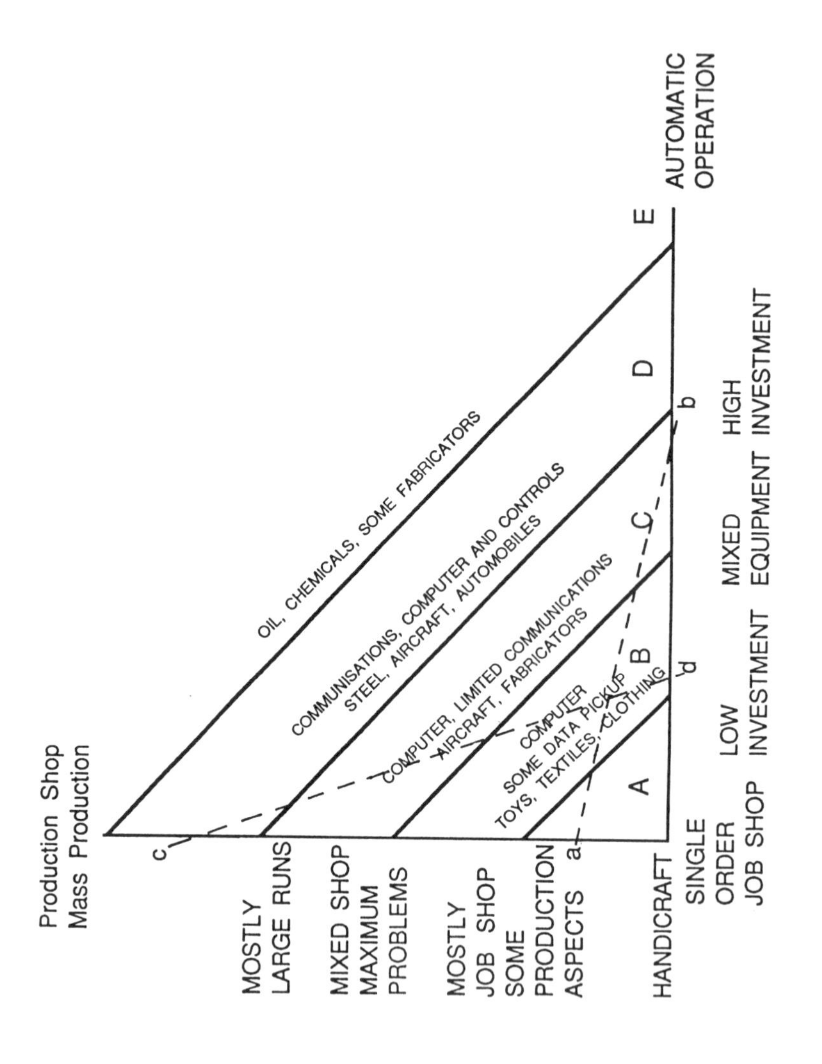

FIGURE 1.2. EQUIPMENT VS TYPE OF PLANT OPERATION

of Plant Operation, relates the industry spectrum to type of plant operation. The vertical axis ranges from job shop operations (low volume--individually prepared products) through production shop operations (high volume--mass produced products). The base lines from the horizontal to vertical axis roughly establish the bounds of computer-peripheral data gathering expectancy.

EQUIPMENT VS. TYPE OF PLANT OPERATION (see Fig. 1.2)

Area A, consisting of individual handicraft items, is not normally interested in data processing; however, specialized programs may be in use on PC's.

Area B is generally interested in small systems with minimum cost data pick-up.

Area C may require a larger computer with limited wired data pick-up.

Area D represents a potential market for overlaying existing investments with control systems involving control computers, data processors and wired data pick-up.

Area E is primarily new construction which maximizes automation at point of construction and is seldom altered after completion. Off-line large scale computing systems are subject to upgrading.

Dotted lines a,b and c,d are drawn to provide a reminder that other combinations exist. There are some low investment high volume operations and some high investment job shop operations. The combination with various business configurations are manifold. This implies that more than one production control and data gathering system may be required within a manufacturing firm. Techniques of control will sometimes shift as the product mix or type of production changes character.

PAYROLL SPECTRUM (See Fig. 1.3)

Typical of the complexities establishing peripheral equipment is the range of problems encountered in payroll accounting. A very simple process can have an extremely complex payroll accounting system. Complexity and difficulty in verifying data range from left to right in Figure 1.3. Most manufacturing operations contain two or more payroll programs (i.e., incentive, daywork, salaries, etc.) and, as a result, require several computer programs to accomplish a payroll. The average apparel payroll system may contain some 12 to 13 classifications of time for the plant payroll. Data gathering requires supervisor coding of worker activity prior to scanning and entering payroll data.

SHOP LOADING (See Fig. 1.4)

Data collection to support shop loading (Fig. 1.4) also varies widely in its requirement. The more detailed the loading, the closer the schedule is timed and the greater the requirement for current information from the shop to the computer system.

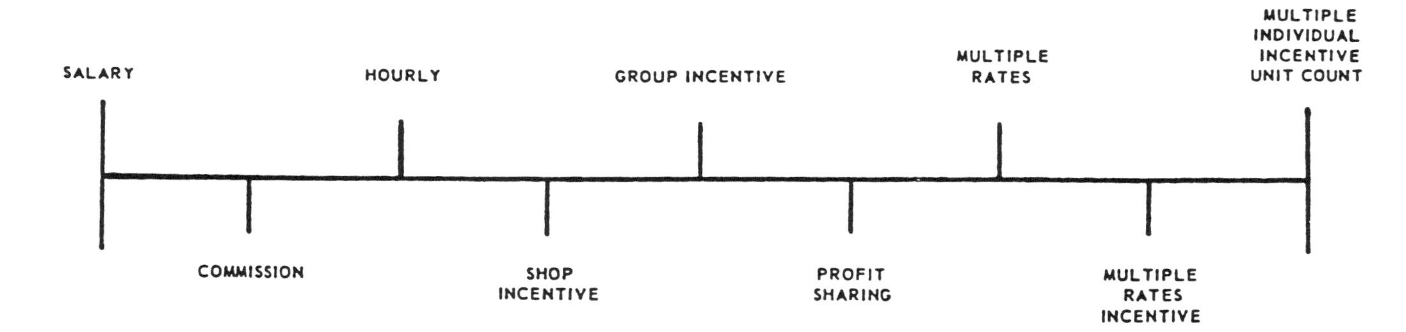

FIGURE 1.3. PAYROLL SPECTRUM

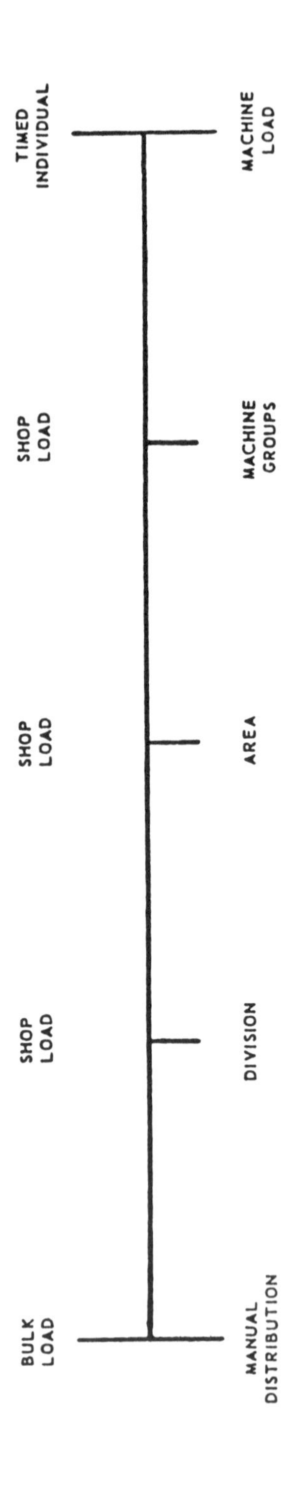

FIGURE 1.4. SHOP LOADING

Similar charts may be drawn for items such as maintenance, inventory control, distribution and allocation of overhead, plant incentive systems for management, and a wide variety of activity normally found within manufacturing enterprises.

Proper selection of data gathering equipment or peripheral computer devices normally requires a balancing of objectives, costs, existing systems and desired controls.

Tradition plays a strong part in the development of manufacturing systems, probably more of a part than in systems within other segments of the economy. This results in frequent illogical requests for data input, processing and reporting methods. Within reasonable limits, these requests must be met if system acceptance is to be secured.

DATA GATHERING FUNCTIONS AND EQUIPMENT

Reporting on the total sum of manufacturing activity provides the basis for abstracting control, accounting and operating data. The principal difference in the three outlined segments of data is in the degree of urgency, not in content. As an example, a worker who becomes ill on the job requires a control decision (replace, halt job, etc.). The worker also poses an accounting problem (hours of sickness, hours at full work, etc.). The same incident becomes part of the operating data (shop productivity, delivery schedule, etc.). In order to make the control decision, prompt notification of proper personnel is required. Data systems should also provide to the foreman or supervisor the impact of the ill person's absence on his or her operation. This information may be: a read-out of scheduled commitments, of alternate workers with some skills, of alternate routings or any number of similar data. Many current data systems have this capability. Training staff in how to use data is as important as the reporting system itself.

Accounting data problems are simpler. A clock-out for time of illness provides the basis for adjusting pay. Time logs are normally provided in pay cycles to permit data accumulation and calculation. Basically, accounting data is batched to provide periodic reports. The data collection system can thereby be less expensive and extensive. However, data must be relayed within stipulated time periods or it is rendered invalid.

The operating data requirements are similar. Most plants operate on a turnover cycle. Productivity may be averaged within the cycle. Since such cycles are based upon prior experience, and prior experience includes illness, variance reports derived from operating data are inclusive of reasonable deviations within given time spans. Again, the data may be accumulated for analysis at a later date. Normally, apparel engineers review portions of this data weekly and consolidate it monthly for manufacturing management. Information technology will tend to displace this activity through an expert system.

For convenience, data gathering problems within manufacturing establishments may be classified arbitrarily into the following broad categories:

a. Control Data - that segment requiring immediate action.

b. Accounting Data - periodic time cycle data.

c. Operating Data - historical analysis for improved control.

Control Data

For the purposes of this analysis, control data is defined as information utilized by management to generate immediate follow-up or anticipatory action. There are two types, short-term spot and long-term management decisions. True short-term control poses extensive on-line data gathering, processing and display problems. Long term management decisions do not require extensive data gathering equipment. Most firms to now have compromised with interval data gathering systems for most decisions (i.e., production bottlenecks) and expect supervisors to handle personnel movement decisions. Information technology is becoming sufficiently low in cost to displace interval data gathering systems.

The extent to which controls are applied within a manufacturing establishment defines the volume of data required to satisfy management. Controls may be applied to one element of production such as key machine, department or section, or may be extended to every element of production. Even within an industry group, plant managers and operating personnel differ as to the extent they require current information from the plant floor. Control data may pertain to the location of work within a production sequence rather than to the elements of production themselves.

In order for control data to be effective, it must be promptly obtained, relayed and acted upon. This requirement provides one of the more difficult and expensive aspects in the development of a data gathering and processing system. In a following chapter, some aspects of the impact of organization upon the flow of data will be covered.

In general, the data gathering and processing problem created by control data requiring prompt and immediate action is solved by use of specialized equipment. Among the items of special equipment used are: industrial control computer systems; data loggers; two-way on-line communications; computer-control systems of various types; electro-mechanical systems; plus various combinations of telephone, intercom and messenger systems. Justification for expenditure in such systems is of necessity related to the size of plant area covered, volume produced and similar criteria. The advent of small less expensive computers has provided the basis for substantially expanded activity in this area.

Specifically, a control computer currently costs from $25-100,000 per station, a data logger from $10-50,000 for the computer. Some smaller units based on PC's are now being offered, ranging from $4-25,000. In many cases, vision systems and robots are being tied to control computers. Justification for such systems is derived from fractional savings on large volumes of products, from unusual profit situations or from multiple product changes.

Electro-mechanical or electronic systems which tie directly to lathes, acetylene torches, drill presses, knitting machines, etc., to count unit output and to display this count at a central station, or on appropriately located video tubes, sell complete for from $200-500 per station. Basically, these systems gather data at the machine level concurrent with activity performed. Manpower is utilized to make decisions. Control to point of operation is exercised by intercom or telephone. A prime purpose of these systems is to remove paperwork from the plant floor. In-process flow control also may be monitored through electronic or electro-mechanical systems.

Data gathering systems which display data gathered from machines or operations on the plant floor through electro-mechanical hookup may be utilized with a computer through relatively inexpensive buffering or through paper tape and manual processing. Some systems use optics to scan this type of data.

The intercom and telephone systems used to report trouble and to institute decisions on a spot basis still remain the basic framework for transmitting and utilizing control data. Wireless phone systems using beepers are also used to facilitate control action.

Computer production control systems attempt to predict or anticipate trouble through improved scheduling and to provide access to alternate information in the event of difficulty. Concurrent or parallel updating of master files is required to accomplish this function effectively. The main frame production control alternate schedule program may be accessible through a distributed process system (with interrogation units). This distribution of control information is currently coming into wider use. Up to now, most data gathering systems, offered in conjunction with a main frame random access unit, function on a historical basis rather than on immediate basis, hence fail to satisfy short term control data requirements. Interrogation was often made to an out of date file. Distributive processing represents an effort to maintain totally current files.

Accounting Data

Accounting data consists of information pertaining to operations required to generate costs, payrolls and to locate investment or expenses. Just as plant operating personnel require varying degrees of information to control plant functions, so also do fiscal personnel manage their data accumulation, budgeting and reporting in dissimilar ways. An accounting data center may or may not coincide with production data centers.

The bulk of the data gathering and computer program systems currently offered for manufacturing purposes were originally developed to satisfy accounting data requirements rather than control or operating data requirements. Recent computer systems place more emphasis on the current use of data for decision making. Justification for accounting data systems was originally rooted in cost reduction for the collection and conversion of data. These older accounting systems pre-supposed a tabulating or computer system in being, and were largely directed toward reducing key punching costs and errors.

Accounting data systems also seek to reduce the cost of timekeeping on the floor through semi-automatic, simplified recording of pertinent data. By including the gathering of operating statistics within these systems a combination of the production control, expediting and timekeeping functions can sometimes be obtained at a net reduction of plant clerical personnel. Certain types of accounting data, particularly in respect to incentive payrolls, split calculations in respect to data gathered by using smaller computers in the plant, then by transmitting data to the home office to calculate gross to net pay.

A substantial portion of the larger scale assembly type manufacturing establishments operate with "zone control" data collection points. A much larger portion of manufacturing establishments are basically "bench controlled". The coupon, ticket and tag systems of Dennison, Kimball, Cummings, etc., are primarily utilized to satisfy bench controlled accounting data problems. These systems shift the point of recording by a timekeeper or zone to the individual worker. Preprinted and/or punched documents control worker reporting accuracy. The preprinted cards, tags, etc., are accumulated by the worker to account for activity. Most apparel firms use this system. Cost per station for a system of this type is absorbed into the worker's time. Ticket systems represent one of the lowest cost per unit data collection systems in current use; however, as a system it lacks flexibility and poses many problems of interpretive analysis.

Optical scanning provides a basis for a combined zone--bench data control system. Present optical-scanning equipment is provided as a computer accessory together with OCRA font or bar code printers. Bar codes with optical fonts have spread from work coupons through inventory and other records.

Optical systems offer a wider range of flexibility than either ticket or wired work station systems because of relatively low costs and the ability to accept variable data beyond the ticket or card limits. However, optical systems have been basically historical data gathering systems. It is only recently that unit processing systems have been adapted to developing control data.

A final system, interrogation and display, mentioned in control data systems, properly belongs in the accounting and operating data systems area. Presently most suppliers of computers are providing random access memory with remote inquiry units. These units price out at from $500 to $1500 per station depending upon the terminal selected. When such units are tied to an adequate data collection system and used in conjunction with a computer which updates files as events occur, the system can become a control data system as well as an accounting and operating data system. Used as an accounting and operating data system, the remote inquiry terminal can generate reasonably timely information, not over 4 to 8 hours old, as to plant or customer in-process status.

However, it is still a batch system rather than a true on-line system.

Operating Data

Operating data is defined as information as to the location of specific customers' orders within a manufacturing operation, maintenance of proper inventory levels, calculation of operating variances, and accumulation of statistics relating to plant performance. Operating data can be as extensive as time-motion studies for every operating unit (or job) or as brief as volume statistics on total plant output versus costs and input. Analysis of operating data can occur daily, weekly, monthly, quarterly, semi-annually and/or annually depending upon the production requirements and plant characteristics.

Decisions deriving from operating data are primarily of a planning nature leading to improved plant productivity through improved scheduling. Cost accounting data is related to, but not necessarily inclusive of, operating data. In a plant operation, control can be exercised over known limited operations using reports deriving from these bottleneck operations to characterize the plant. In such a case, accounting data collection problems would be more extensive than operating data collection problems. Control decisions based upon control data could occur in areas other than the bottleneck areas, hence a communication network would have to exist for these purposes irrespective of accounting and operating data requirements.

A data collection system which gathers all data, control, accounting and operating data and satisfies all timing requirements for each use is of necessity on-line with everything in the plant. Generally speaking, systems of this type are only now becoming affordable.

Operating data systems are largely manual in present plant practice. Among such systems are:

a. Ditto routing sheets annotated with station by station activity.

b. Master routing sheets updated by clipped on, manually prepared work slips forwarded to production control centers.

c. Prepunched tab cards annotated with station activity and accumulated for later punching and transfer to activity reports.

d. Foreman, leading man and worker private logs of volume activity, reviewed by management through manually prepared activity reports.

e. Work station wired systems.

f. Tags, tickets, slips, etc., systems.

g. Data logger and industrial computer systems.

ON-LINE SYSTEMS

Reporting cycles for operating data are more closely related to equipment maintenance cycles than to calendar division. Data on excess downtime, variance in scrap loss, etc., does not necessarily coincide with the reporting cycles established for fiscal purposes.

In some respects operating data may represent a continuous information flow to plant management since machine cycles, operating characteristics and similar data seldom coincide from center to center or machine to machine. This characteristic of operating data poses data processing problems which are generally obscured or ignored. In submerging operating data reporting into an arbitrary calendar division of time, the value of such data is often diluted.

The installation of data collecting systems have tended to duplicate or parallel manual collection systems. In general, the systems covered in the two previous sections have been utilized for the purpose of generating operating data. The economics of such installations generally reduce their scope and their effectiveness. Data is gathered for "must" purposes; i.e., payrolls, inventory, cost allocation, with secondary consideration for operating purposes.

CONCLUSION

Serving the data gathering problems associated with manufacturing operations requires a detailed understanding of:

a. Information technology.
b. Type of industry.
c. Product mix.
d. Level of production.

e. Controls in use.
f. Accounting systems.
g. Economic base.

Generalizations are dangerous. Pure logic alone will not insure a successful system. The secret of making a successful installation is the obtaining of a balance between gathering too much data and too little.

2

Apparel Computer Overview

INTRODUCTION

It seems appropriate to include a brief summary of computer applications within the apparel industry--and the current general trends before going into more specific detail on the use of computers.

Before the apparel industry realized it, dependence on electronic computer technology had become an integral part of its way of life. If the computers were suddenly to stop functioning, the number of businesses that would come to a screeching halt and the degree of disruption to the industry would be significant.

In this age of information, the apparel industry also has experienced a steadily increasing proliferation of computers designed to help manage business and compete within the marketplace more effectively. To understand the scope of applications for which computers are or can be used in the apparel industry, we will review the history of computers, explore the evolution of their use in the apparel industry, and consider many current applications of computers in the manufacture of apparel.

Computer technology and its almost universal acceptance have opened new areas for potential application in the apparel industry. These areas will be explored as well as trends for future use and applications development resulting from this new technology.

Defining the term "computer" and discussing its development will provide a perspective for this discussion. Webster broadly defines a computer as a device or person which determines or calculates, especially by mathematical means, or more specifically as a programmable electronic device that can store, retrieve, and process data. With this definition in mind, we will begin a review of the history of computers.

HISTORY OF COMPUTERS

If you look at computers in their broadest sense, you will find that man has been concerned with the problem of developing a mechanical computing device for thousands of years. The abacus, whose origins extend back approximately 5000 years, existed in the Tigris-Euphrates Valley in southwestern Asia Minor. The Chinese abacus, another version of this device, was invented in approximately 2600 B.C. This device and a similarly used Japanese device, the soroban, are still used today with incredible speed and accuracy. In fact, various forms of the abacus have existed in all ancient civilizations as an aid in calculating numbers.

Calculators also existed in medieval England. Two thousand years before the Middle Ages, monuments known as Stonehenge were built in England. Today many archeologists believe these stones were used as an astronomical computer.

During the Middle Ages and the Renaissance, many people worked on the problems of computing, including Leonardo da Vinci who drew designs for a machine to solve computational problems.

In the early 1600s, the Pascaline device was developed by Blaise Pascal. It is best known to modern data processors as the first calculator. This calculator performed addition and subtraction processes essentially like calculators still in use only a few decades ago. (A computer language is named after Pascal.)

Following the Pascaline device, another major development in computers occurred in the late 17th century when Gottfried Wilhelm Leibniz developed a machine that allowed the user not only to add and subtract, but also to multiply, divide, and extract square roots. Machines built using principles Leibniz developed were in use through the 1930s, and another type of machine based on his principles was still in use until electronic calculators were developed.

In the 1800s, work continued on the development of a computing machine. Charles Babbage, an Englishman born in 1791, spent most of his life working on the development of a mechanical computer and developed several different machines. One of them was called the Difference Engine. A later development was the Analytical Engine which could perform mechanically most of the features of today's digital computers.

Another man who played a major part in the development of computers during the 1800s was an English mathematician named George Boole. The publication of his major work, An Investigation of the Laws of Thought, proved to be the basis for what we know today as Information Theory. The theory of logic which was developed in his Information Theory became the basis for the development of telephone circuit switching and the design of electronic computers.

One last person should be mentioned in tracing the evolution of the electronic computer. Based on Boolean algebra, a type of algebra developed by George Boole, an American--C. E. Shannon--was able to develop a universal measure and universal unit of information--the binary digit, or bit. This measure is the key to modern electronic computers.

This brief historical survey of computing illustrates the long history of man's search for a means of performing calculations and manipulating data by machine.

In understanding today's computers, however, we can begin with the advent of electronic data processing in the 1940s. During this time, the U. S. Army developed the first computer known as ENIAC which is an acronym for Electronic Numerical Integrator and Calculator. Electronic circuitry in this machine eliminated the use of mechanical parts and produced extremely rapid, for then, calculating speeds.

Also during the 1940s, computers were developed which were able to perform calculations on their own using instructions stored in memory. These first electronic computers used vacuum tubes. They generated extreme amounts of heat, and a great deal of space was required to accommodate them. The RCA Bismac, installed in the Detroit Arsenal, was built in arches so that you could walk through the machine to locate burned out tubes.

During the 1950s, a new breed of computers was developed using solid state or semiconductor technology. These computers provided much larger storage capacity and faster calculating times. With the advent of the second generation computer, applications in government, research, and business increased greatly.

Following these computers, a third generation of computers was developed. Using a silicone chip or integrated circuit, the storage capacity of computers was expanded exponentially. The cost of computers has steadily decreased. In the two decades from 1960 to 1980, the cost of storing and using a given amount of information decreased by a factor of 100,000 to 1. In addition, the decrease in size of computers resulting from new space age technology has enabled the computer to operate in a much smaller environment.

The proliferation of computers today has provided us with a vast range of systems. The microcomputer, a desktop device which enables a user to store, retrieve, and manipulate data, is the smallest of these. The next largest is a minicomputer. Minis basically are low-cost computers which provide many of the capabilities of the large, general purpose mainframe computers. The upper end of the computer hierarchy is the mainframe. This is the largest, most versatile type of computer which performs very complex operations and allows many tasks to be performed simultaneously within the system. In the past few years the distinction between the minicomputer and the mainframe has gradually blurred, blending the upper end of the minicomputer market and the lower end of the mainframe market. There is also a group of high speed scientific computers which perform complex calculations almost instantaneously.

In addition to the technological advances which have enabled computers to become smaller and cheaper, the method of processing has changed over the past 10 to 20 years. Initially, inputting and processing data were done in a batch mode. Batch processing provides for the accumulation of data over a period of time and the processing of that data at one time to update files in the system. Drawbacks of this type of processing include delays in providing the user with updated information and the possibility of not getting certain data into the system because of errors in the input.

As technology has improved, interactive processing has been developed and is replacing batch processing in many environments. Interactive processing allows the user to enter data directly, to receive error messages as data is being entered, and to see immediately the impact this information has on information in the files. This type of system provides much quicker response to information and tends to eliminate error recycling (which causes delays in getting the information into the system). It also allows the user to interface directly with the computer rather than through a keypunch operator.

Two other terms--software and hardware--should be defined at this point. "Software" refers to the instructions or programs which tell the computer what to do with data and how to operate. "Hardware" refers to the machinery itself. All of the computer's components--the processor which manipulates the data, the storage media including tape and disk drives, cathode ray tubes or screens, and all type of printers--are included in the category of hardware.

DEVELOPMENT OF COMPUTERS IN THE APPAREL INDUSTRY

Computers have been used in the apparel industry for many years. Larger companies began to use these machines in the early 1950s, particularly for office functions such as accounting, billing, order entry and processing, and payroll. These computers, utilizing punch cards and batch processing, did not have the power or flexibility of today's machines. To some extent the first apparel computers were status symbols rather than useful machines.

Later, as computer technology advanced and as computers became cheaper and more necessary for successful competition in the apparel business environment, more companies began to acquire computers.

Slowly, more sophisticated systems were designed which integrated information from various corporate activities into a total company system. Even as recently as 5 or 6 years ago, however, many medium-sized companies still did not have a computer because of the prohibitive cost of maintaining a computer system in terms of manpower, hardware and software.

With the development of minis, however, this situation has changed dramatically and many companies now find it quite justifiable to purchase a minicomputer which performs accounting, merchandising, and production control applications.

Data processing timesharing companies rent time. This allows firms that cannot afford computers of their own to rent time to run their programs or operations on an outside computer. Companies can take advantage of the positive aspects of electronic data processing without large investments in hardware or programmers.

Another method of using a computer system without a major company investment is the use of a service bureau, a company that runs clients' programs for a fee. Through this method the customer does not have to purchase computer equipment for his premises, although a remote terminal is usually necessary to operate programs from his plant.

After office functions, inventory applications were developed in many companies. Often, the first type of inventory installed was piece goods, which for many was easier to develop in a batch environment than in-process or finished goods inventory systems. When interactive systems became more feasible, computerization of both in-process and finished goods inventory followed. However, some companies have maintained a computerized in-process and finished goods inventory using a batch system.

The development of forecasting systems, which attempt to predict the success of styles in the marketplace, often followed the computerization of order entry and inventory. With these applications automated, many companies then were able to move into the production control area taking orders and forecasts for orders, comparing them to inventory positions, and generating production plans.

As applications were developed in the apparel industry, companies often developed each application as a discreet function with duplicated data bases. Later, consolidation and integration of these functions into management information systems ensued. Applications originally developed for batch processing are being or have been converted to on-line or interactive processing.

The advent of the minicomputer has involved managers much more in the use of systems and in the decision-making process. Production control and scheduling are two examples. Before computerization, these functions required very intensive clerical support and substantial "number crunching" to produce production requirements. Managers often spent all of their time manipulating figures rather than reviewing information and making management decisions. With the advent of the computer, this function has changed dramatically.

During the 1960s, the use of computers in the apparel industry expanded greatly. Computerized pattern grading and marker making systems were developed about 1965. These systems utilize the ability of the computer to perform rapid calculations with consistent accuracy and also to store large amounts of data in memory to grade ranges of patterns very quickly from one initial pattern input. The graded patterns are then stored in memory and the computer is used to generate markers for cutting. Today, computer-assisted grading and marking are widely used in the apparel industry. It is expected that this trend will continue and that more and more companies will be moving in this direction.

Computerized cutting followed the introduction of computerized marking and grading. This cutting is accomplished by using the computer to drive a cutting knife through spreads of fabric. Cutting is also done by computers guiding a water jet, plasma or laser. Because of the accuracy, speed, and flexibility of automated cutting, substantial savings have been realized by users of these machines. Savings in piece goods inventory also have resulted from the use of automated cutting systems.

In addition to computerization in the cutting and cut planning areas, the use of microprocessors based on semiconductors and programmable chips has become more widespread in the sewing area over the past 3 to 4 years. The Technical Advisory Committee of the American Apparel

Manufacturers Association produced a report in 1978 on three computerized sewing and pressing machines. By the summer of 1980, at least 15 different examples of this type of equipment were found on the market, with many more in the process of development.

To recap the growth of computers in the apparel industry, we see that computers began as very large, powerful tools for manipulating data. Initially, computers were confined primarily to office use, but as technology grew, they became smaller and cheaper which allowed for more varied use. Today, computers are being used as production aids in addition to their function as information systems.

COMPUTER APPLICATIONS

While apparel companies have organizations like the rest of industry, some of the function terminology is different. Many of the functions which take place within an apparel company (Fig. 2.1) have information needs which can be supplied by data stored in a computer. As a matter of fact, much of this information can be generated from only a few carefully designed documents.

To understand these documents and how the computer system would manipulate the data provided by them, we will examine each functional area and discuss briefly some types of information generally needed for that area.

Sales

Salesman Performance Report: The salesman performance report identifies the

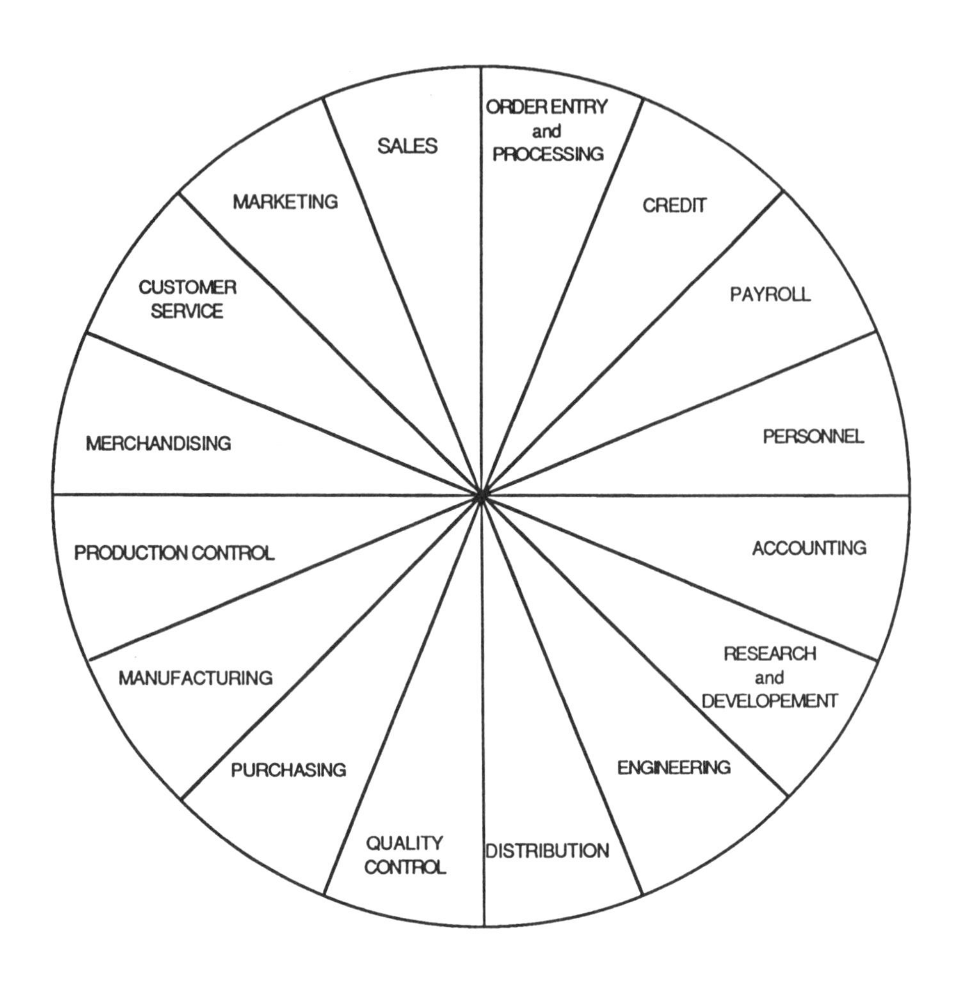

FIGURE 2.1. FUNCTIONS WITHIN AN APPAREL COMPANY

performance of a salesman against established goals for his/her area and shows information relating to historical sales as well as current sales in both units and dollars. This type of report allows the salesman and his manager to determine whether the salesman is meeting his objectives.

Commission Report: The commission report provides information regarding commissions earned by each salesman. Often these reports show goals as well as prior seasons' performances.

Customer History: When salesmen call on customers, they should have information regarding each customer's past buying habits. The customer history report shows which styles a customer previously has bought, the historical dollar volume the customer's purchases have represented, and other similar information.

Other Reports: Other reports of interest to the sales department include reports on sales by product and sales by geographical area. These reports enable sales managers to predict and forecast future sales performances.

Sales forecasting and monitoring of these forecasts are important functions in the apparel industry, not only for financial planning but also orders and are maintained by the computer as well as reports which age accounts receivable.

Payroll

Payroll information is often produced or maintained on the computer. Information in this area includes employee payroll history, employee benefits contribution history, employee performance reporting, tax records, and work-in-process reporting. Executive payrolls may be placed on a service bureau computer to maintain confidentiality.

Personnel

Personnel functions may be aided through the use of a computer by establishing an employee history on the computer. Computers can also produce labor reports, such as absenteeism and turnover. Employee benefits can be monitored, thereby reducing clerical requirements and simplifying operations. Through the use of the computer, criteria may be established to alert management when employees are eligible for company benefits.

Accounting

The accounting functions in most cases are supported completely by the computer. General ledger, accounts payable, journal entries, and maintenance of these entries are all done on the computer. Analysis of various accounts as well as quarterly and annual tax reports can and should be produced by the computer.

Research and Development

Many companies have a research and development department. Information used by this area includes inventory of sample items, costing information regarding new products being developed, pattern design functions which enable this department to experiment with new patterns, and test marking which facilitates costing of new samples. Research and development in the apparel industry seldom develops new equipment or manufacturing systems.

Engineering

Applications that might be required by engineering include product costing, master operations information, style bulletins, cash flow projections, investment analyses, return on investment analyses, lease or buy decision-making problems, manufacturing performance analyses, and development of training curves for operators. All of these functions can be supported by and maintained on a computer.

Distribution

The distribution center or distribution function in a company requires information regarding finished goods inventory, unshipped orders, shipping manifests and shipping documents, routing information, stock locator information, traffic assignments, the production of billing information, and the ability to enter information regarding returned items.

Quality Control

Quality control accumulates a very large volume of data concerning operations in a plant. This information is used to produce the following types of reports: work-in-process inspection reports, operator quality performance reports, final audit reports, reports describing and analyzing repairs, percentage defects in sewn and cut work as well as raw materials, fabric inspection reporting, and shading. Not only is information regarding shading generally maintained in inventory records, but currently computers are available which actually shade goods. Direct input of other in-process quality control data is gradually increasing.

Purchasing

Information used by the purchasing function includes piece goods inventory information, trim inventory, and net position reporting. This information is then used in material requirements planning. Computers today consolidate information from these applications, and produce reports which are used by the purchasing area to determine ordering requirements. Purchase orders are then generated and open purchase order reports are produced.

Manufacturing

The manufacturing function can be divided into three areas: sewing, marking, and cutting. One report used by sewing supervisors is a labor performance analysis which enables supervisors to determine operator efficiency. From this type of information and from production and delivery requirements, it is possible to balance work-in-process in a plant. This type of report, along with work-in-process controls, enables management to determine its position each day. In the sewing area, sewing and pressing equipment using microprocessors to reduce operator skill requirements has significantly increased productivity and quality.

The level of sophistication of the computerized equipment recently developed to produce markers has increased over the past ten years. Software is now available which will handle cut planning by grouping requirements in the most efficient manner to generate a cut. With this equipment markers are made and stored within the system and can be used many times without having to re-mark.

In addition, pattern grading and pattern design are functions of these systems. The pattern design software available today is a computer-based interactive graphics system which automates the process of making design variations in block patterns. This package can also be used as an alteration tool for changing the shape of production patterns when used in conjunction with an automated marking system. The system has been used in the slacks and shirt industries and in women's wear, but has not yet been used extensively in the men's tailored clothing segment of the apparel industry.

Automated cutting systems allow spreads to be cut based on a marker stored in computer memory. Several types of these automated cutting systems exist. A cutting system using a blade which cuts through fabric held firmly under vacuum pressure is the most widely used. Another type of automated cutter is the water jet cutter which uses a stream of water under very high pressure to cut through material. A third type is the laser cutter which uses a laser beam in a manner similar to the water jet. The latest displayed cutter uses plasma, an ionized gas, to cut.

Production Control

Production control, which is usually responsible for scheduling and expediting work through the manufacturing process, requires certain information in order to maintain its function. Among these requirements are forecast monitoring capabilities. The computer can monitor forecasts against orders as they are entered and report deviations from the original forecast, allowing for changes in the original. Another type of system used by the production control area is plant loading. With this system, management enters requirements and the computer assigns them to certain plants based on specified criteria. As these requirements are loaded, comparison with plant capacity is made and over/under capacity situations are rectified.

Production scheduling provides for slotting in production to maximize manufacturing efficiency and to produce finished products which meet delivery schedules.

But planning is another production control activity performed by computers. In a cut planning system, requirements for production are grouped to produce the most efficient cutting combinations which meet sales requirements.

Work-in-process monitoring and control provide management with information regarding where units are in the production cycle based on information supplied by operators. The same information which produces payroll reports can be used to produce work-in-process control information.

Inventory control is another critical area to production operations. Inventory control for all types of inventory can be maintained by the computer. In addition, material requirements planning used inventory information and production requirements to generate net position reports for each trim item and fabric used. It also incorporates an open purchase order system which shows management which trim and fabric requirements are on order.

Merchandising

Merchandising uses information to help determine the type of product to sell and manufacture as well as the makeup of that product. Style history reporting provides information regarding the types of styles that have sold within the target market of the company. Vendor reports provide information relating to the type of goods that various vendors supply, the reliability of those vendors, and information regarding lead times, etc. Sample inventory information as well as regular inventory information may be helpful to merchandising and also should be on the computer. Often, in trying to fabricate lines, merchandising tries to utilize fabric left in inventory from previous seasons, and access to inventory information becomes invaluable.

Customer Service

Customer service provides the link between sales, the customer, and the production control area. In order to service customers, report order statuses, and project deliveries, customer service needs information regarding orders and stock. This information is provided by the computer. Order history is also required by customer service to deal with customer problems. Customer service very often enters orders for the company or prepares them for entry. To do this, customer service has to be familiar with the requirements of order entry--the parameters and edits which ensure that orders are accepted in a timely manner.

Inventory is another area to which customer service must have access in order to take orders over the telephone. By having access to inventory on the computer, customer service is aware of availability as customers make inquiries. To accept orders, customer service often must refer to credit history information which should be available on the computer as well. Some computers use "split screens" or "windows" to display multiple data.

Marketing

Marketing may be the responsibility of a marketing department or may be handled by other departments within an apparel firm. The purpose of the marketing function is to provide direction for the company in terms of which markets it should be developing and selling and which types of products it should be producing. To do this successfully, access to information regarding sales and customers as well as demographics within various markets is important. In addition, sales forecasting is very often monitored by the marketing area to plan new strategies successfully.

THE COMPUTER SYSTEM AS A FUNCTIONAL INTEGRATOR

Order Processing

To supplement this overview of computer applications in various functional areas within an apparel company, we will examine a major process within an apparel company--the order flow--and trace the activities and the functions which are affected by this order is processed. Figure 2.2 illustrates this process and the areas involved.

The receipt of an order can occur either in the company facility or at a remote location, i.e., the company showroom, or the customer's store. Today, an order can be entered at any remote location using a portable terminal, and through the use of telephone lines the order can be transmitted directly into the computer system of a company. If the order is entered in the company, the order would be edited and checked visually before being entered into the system. Once that occurs, the order would be entered and computer edits would be performed to ensure that information on the order is correct.

The functional areas which have been involved with the order up to this point are illustrated in Figure 2.2. First, the sales staff has been involved either directly or indirectly in initiating the order. Once this order is written, it is sent to either a customer service or order entry department within the company itself. These functional areas are then responsible for editing and entering the order. After the order has been edited and entered into the system, questions regarding the status of the order, delivery of the merchandise, and its impact on production can be answered.

A credit check must be performed before an order is scheduled for delivery to determine if that customer's credit

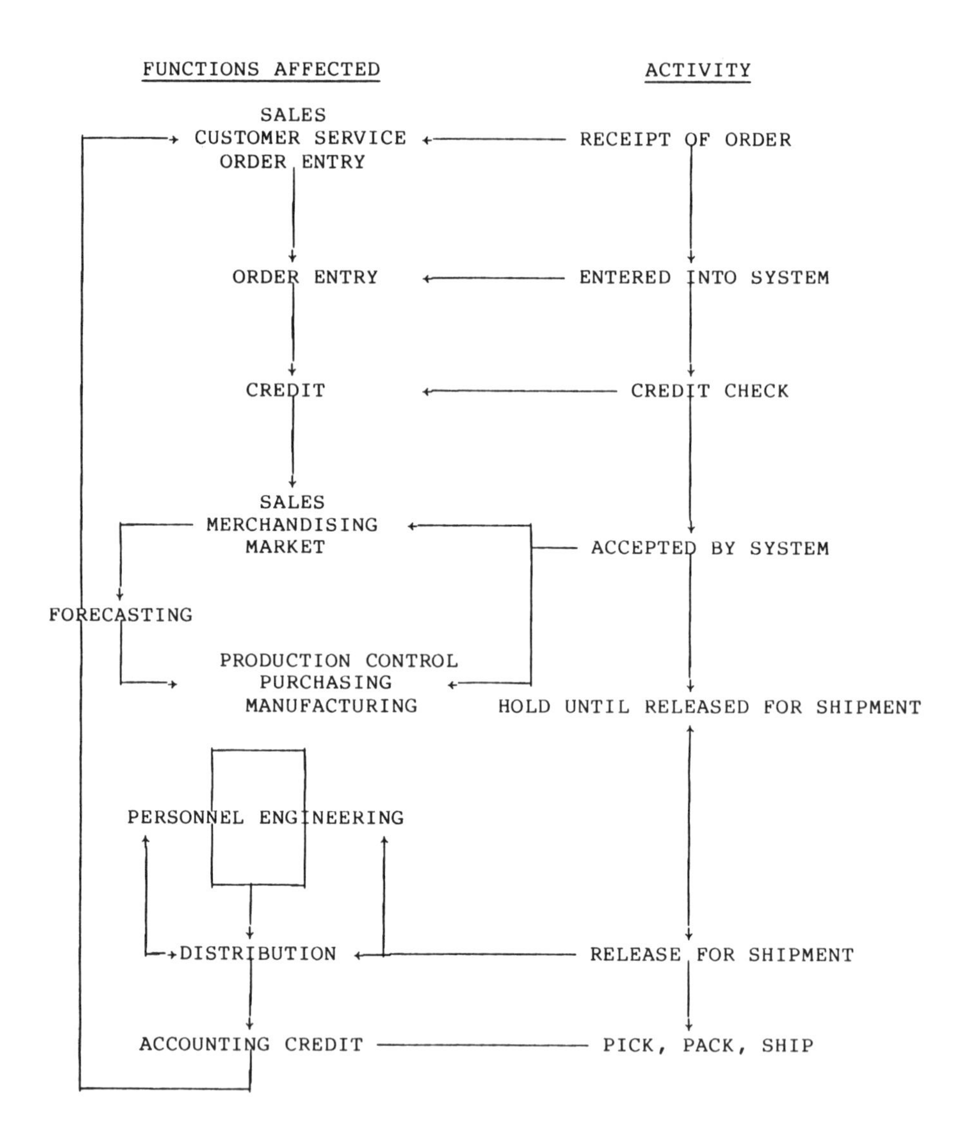

FIGURE 2.2. ORDER FLOW

limit has been exceeded. If it has been exceeded, the order will go no further in the process. If it has not been exceeded, the order will be processed into the system and will impact several other functions. Therefore, the credit department becomes involved at this point and the order is posted to the credit screen. Once favorable credit screening has occurred, the order is accepted into the system and several different functions are initiated. First, this order impacts the sales, merchandising, and marketing areas which monitor performance by style, customer, and season. Also, production control, purchasing, and manufacturing are impacted by this order either directly or through the forecasting system.

Production in an apparel company is geared to respond to one of two basic inputs. A company cuts and manufactures against either orders or a forecast of orders. In any event, orders or forecasts of these orders will precipitate certain activities within the purchasing, manufacturing, and production control functional areas.

In order to produce a style, purchasing must have ordered and ensured delivery of the necessary materials. Production control must have scheduled the work into the plants and ensured that other functional areas involved in the process of producing this product are aware of its requirements. Manufacturing must have ensured that capacity, labor, and machinery are available to produce the necessary requirements. In accomplishing these tasks, the personnel and engineering departments have become involved in staffing, ensuring that equipment is available, and developing rates.

After the goods to fill an order have been made and are in the warehouse, the order may be released for shipment to the customer. At this point, the distribution center becomes involved in the order. Distribution must have the staff to pick and pack the order and be aware of where and when the order is to be shipped to meet the customer's requirements.

Shipping the order creates transactions in both the accounting and the credit functional areas. Accounting generates accounts receivable and general ledger transactions; credit updates the customer's credit record. Of course, customer service must be made aware that this order has been shipped in order to respond to customer inquiries.

The scenario described above demonstrates how a process initiated by an order affects many activities within a company and impacts a variety of functional areas. By using computers to process this order, reports and information screens can be produced which inform all areas of the company of requirements for producing and shipping. Without a computer system to handle these functions, communications between the areas become much more difficult. An order requires the involvement of many different areas to ensure that it is produced and shipped within the required time frame. To ensure that the company is paid for that shipment, other areas also become involved. What is in reality a very complex network of communications can be simplified from the user's standpoint with a computer system.

Style Adoption

To illustrate this point further, another process within the apparel industry, the adoption and processing of a style through the various functions required to produce it, is illustrated in Figure 2.3. The figure is a very simplified representation of the activities and functions which occur prior to and during the season. The first activity, Make Prototype Garment, requires many hours of styling by the designer to conceive and render the original patterns and garments. Figure 2.3 shows the functional areas affected by this activity as research and development and design. In a company which maintains an automated pattern design system, the design of new models might be simplified by the capability to generate on a CRT alterations to existing patterns. However, most of the work up to this point is done manually rather than on a computer. (This course briefly covers pattern alteration using the Lectra computer.)

Following this process, prototypes are adopted by a company management group. Once these prototypes are adopted, they are prepared to be offered for sale. Involved in this function usually are the merchandising, design, sales, marketing, and production control functions. Again, this process is basically manual; however, reports which these functional areas receive regarding sales and style history are helpful in determining what the product mix of the line should be in terms of price point and fashion direction. Following the adoption of prototypes, several activities must occur to produce a style. Pattern grading must occur so that this style can be produced in the range of sizes to be

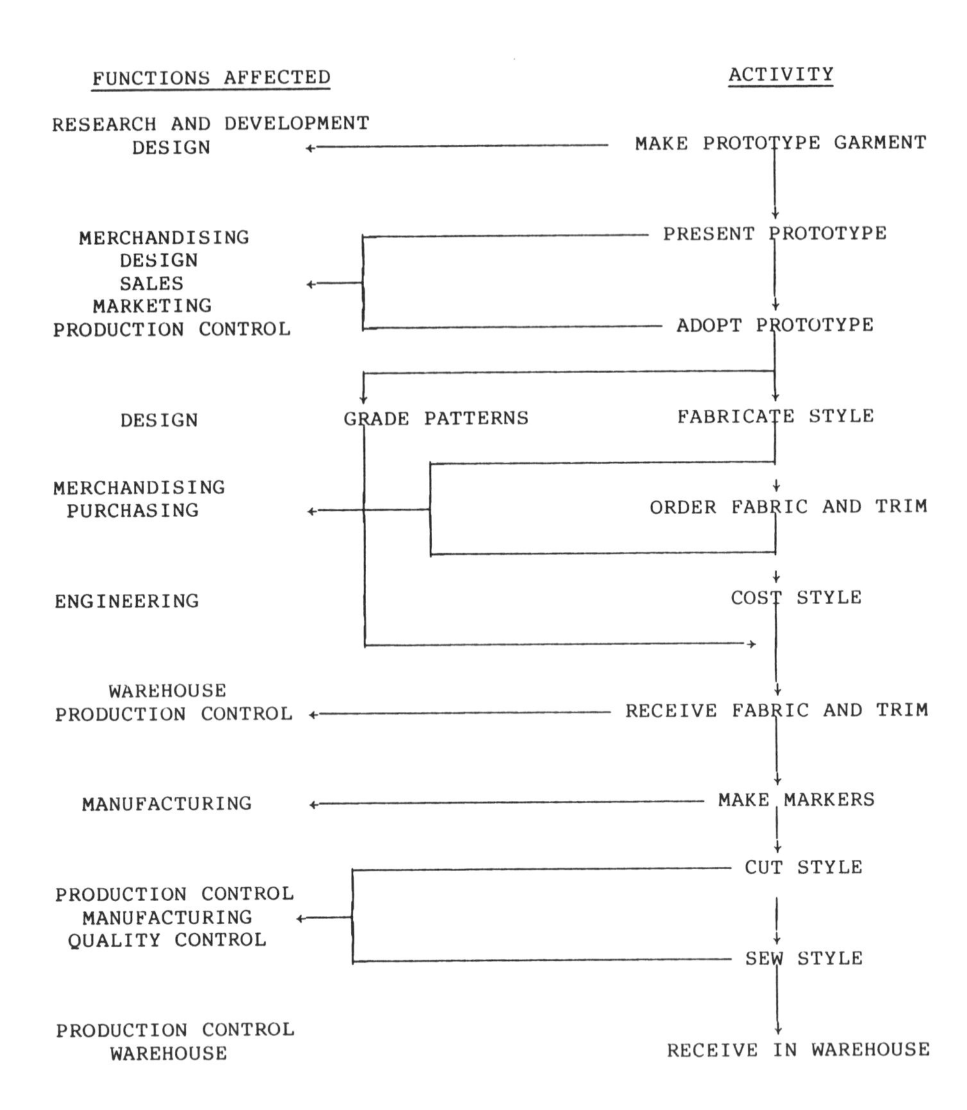

FIGURE 2.3. STYLE ADOPTION

offered. Traditionally, this has been a function of skilled pattern graders in the apparel industry. However, with the introduction of automatic marking systems, these computer-driven systems have the ability to grade single size patterns according to certain rules into all the sizes being offered. Prototype single size miniature patterns may be sent by VeriFax or electronic mail to a computer grading center for sizing.

At the present time, only large companies can take advantage of this function in-house. However, smaller, less expensive systems to perform automated marking and grading are now being introduced. This development should have an important impact on the apparel industry by allowing smaller companies to utilize advanced design techniques.

In addition to grading patterns, the line must be merchandised, trim items must be fabricated and assigned to styles, and requirements for production must be ordered. Usually, these functions are performed by merchandising and purchasing. Master files with information regarding fabric and trim items for each style are developed and maintained on the computer. Trim and fabric are ordered, and purchase orders are held in the system until the required items are received. Production control, merchandising, purchasing, and accounting have access to all information on the computer regarding these purchases. Exception reports are generated when items are late or not received in ordered quantities.

Another function which must occur before styles can be produced is product costing. Costing requires that determinations be made regarding fabric and trim usage and labor content. Costing also requires overhead to be allocated, a function for which engineering is responsible. Through the development of a master operations file, labor costing is greatly facilitated. From this master file, operations can be selected to produce a style bulletin when a new style is adopted. This file and the style bulletin are also used to generate bundle tickets for work-in-process monitoring and control.

Once these functions have been performed and the trim and fabric are received for these styles, the inventory system is updated. Warehouse, production control, and purchasing personnel are made aware of receipts, and production control then issues cuts.

Markers must be made once cuts are issued. Manufacturing or cut planning takes these requirements and generates markers using an interactive graphics scope on an automated marking system. After markers are produced, the style is cut and sewn. As this style moves through the manufacturing process, monitoring of the style takes place utilizing bundle tickets produced by the computer. Information for these bundle tickets is generated by engineering for costing when the style was adopted--and from information entered into the system regarding cuts.

Quality control monitors the quality of work going through manufacturing and reports any defects discovered. This information is accumulated daily and is used to produce daily and weekly quality reports. The computer, rather than quality control auditors, processes the inspection data to produce the required reports.

When production of the style is completed, goods are transferred to the warehouse. Through the in-process monitoring system which has tracked work through the production cycle, production control and the warehouse are notified.

From the preceding overview of the production functions and activities through which a style goes once it is adopted, we can see how the various functions interrelate and how the use of a computer system to monitor and control this work simplifies the communications between the various areas involved.

It is readily apparent that extraordinary opportunities exist to improve communications and efficiency by utilizing a well-integrated, computerized information system. Many of these systems are available today for a reasonable and justifiable cost. Although certain applications can be justified only in larger companies, the most critical operations can be installed in the smaller company today with relatively low-cost minicomputers.

COMPUTER TRENDS IN THE APPAREL INDUSTRY

How available are the software packages needed to meet management information needs in the apparel industry? At present, much of the software designed and marketed to the apparel industry deals with either merchandising, order entry and processing, shipping, accounting and/or credit functions. The wide diversity of different types of manufacturing requirements means that software for manufacturing control, production planning, and generation of cuts has not been readily available in the form of packages. In addition, the development of systems to handle these manufacturing needs most often has been in larger companies which have more difficulty monitoring and controlling these functions manually. However, with the broader use of the minicomputer and the microcomputer, managers in smaller firms increasingly will turn to the computer to handle the production loading and planning functions. Also, as more small companies begin to utilize the computer to perform these functions, the demand for software in package form will be met. Therefore, it is probable that more firms either will develop software in-house or purchase software to handle manufacturing planning and control.

The increasing cost of labor and the decreasing cost of hardware will encourage more companies to acquire microcomputers and minicomputers. As more companies within a particular market move toward computerizing various functions, other companies within that same market will find that they no longer can compete successfully without the installation of computers to handle their information. Because computers provide speed, versatility, and flexibility, those companies which do not have automated capabilities find themselves in situations in which they cannot react quickly enough to compete successfully.

The introduction and acceptance of the microcomputer as a personal computer which can be programmed to meet any number of specific problems does affect the way managers perform their jobs. As this type of computer becomes less expensive, its use will be expanded in much the same manner as the electronic calculator. It will become a tool for handling many short-term problems, and software will be developed which will enable it to be used in many different ways for problem solving. The computer's use for personal memos, communications, and other functions should increase dramatically over the next several years.

The use of computer graphic capabilities (i.e., McIntosh) will continue to increase. This will have an effect on the way management uses information and the way information is presented to others. By using graphic capabilities, trends and historical data become more meaningful to many who currently do not understand the presentation of information in numerical reports. The use of computer graphics will also continue to expand in marker making and the pattern design areas. Currently five firms are working on 3D design in color which can be converted to two dimensional patterns for cutting. The cost of this technology will continue to decrease, allowing smaller companies to afford this capability.

In terms of the effect of computers on direct labor operations, automation through the use of microprocessors for various job elements will continue. Automation will continue to deskill jobs. This will reduce job labor content and make the manufacturing process more productive.

An area which is presently experiencing incredible growth is robotics. Industries are beginning to utilize robots to perform many tasks that are dangerous, tedious, and/or boring to human labor. Today's robots can be programmed to perform a variety of unrelated tasks. Robotics will extend labor's ability to perform, thereby reducing labor costs and improving productivity in many apparel industry areas. Robots greatest attribute is flexibility, allowing changes from season to season as styles change and the transfer from one job to the next as the product mix changes.

The main drawback of robotics in the apparel industry at the present time is the inability of robots to handle limp fabric. Once this problem is overcome, perhaps through the use of "smart" robots which can see, differentiate, and manipulate various fabrics, the use of robotics in the apparel industry will increase. Presently firms are researching the possibility of robots "seeing" fabrics and "selecting" colors. Applications which have potential for future apparel use include:

a. Unloading of fabric bundles
b. Die cutting
c. Operating small tools
d. Transporting packages
e. Bonding
f. Folding
g. Loading profile stitchers

In the next decade, opportunities for dramatic increases in productivity are possible through the use of computers and microprocessers and the use of robots to perform many tedious tasks. The potential for changes in sewing manufacturing is tremendous. As technology grows in these areas, changes will occur in the apparel industry, possibly making manufacturing in our country more competitive with countries whose labor is many times cheaper than ours.

Electronic communication is another area with rapidly developing technological changes. One very important way in which electronic communication affects the apparel industry is direct transmission of orders from major chains to manufacturing faciliites, creating a much reduced in-process time while eliminating the need for mailing orders. Another area of impact is the linkage between fabric production and apparel manufacturer. Potential for inventory reductions as well as more responsive meeting of market demands is possible using direct communication between vendor and manufacturer. As a matter of fact, with large chains, it is becoming almost mandatory to be able to handle this type of electronic ordering in order to do business.

Technology is currently moving this process one step closer to the consumer. Standard coding is now capturing data directly from the retailer's cash registers and translating that into sales information by product categories to enable stores to reorder using this information. Store computers are now linked to apparel firms' units to initiate resupply orders. This quick turnaround response to what is selling in stores today, as now transmitted to the manufacturer, will allow the retailer to reduce the production of poorly selling goods and divert manufacturing capabilities to styles which sell better. The same action is reflecting back to raw material/textile firms who are tying to apparel computers. This business environment creates the possibility for much larger profit potential and reduction of inventories at all levels which do not sell. The need for computers to handle large volumes of data quickly is increased. However, without this capability, firms cannot compete in the electronic environment.

SUMMARY

To summarize, then, the following trends seem to be prevalent in the use of computers in the apparel industry.

a. Greater availability of production control and production planning software.

b. Expanded use of the microcomputer by managers as a personal aid to accomplish their jobs.

c. Greater use of graphic capabilities for reporting as well as design functions.

d. Increased reliance on the computer to simplify jobs requiring high skill levels in the industry.

e. The use of robotics to eliminate many tedious, boring, or dangerous jobs from the manufacturing functions in the apparel industry.

When all of these changes are viewed as a whole, the potential for increased productivity and better response to the marketplace is impressive. All of the apparel industry will have to work to meet the challenge of this increased technology. However, once met, this technology provides the industry with increasing opportunities for growth and new challenges for tomorrow.

3

Organization Functions and Information Flow

GENERAL

Manufacturing firms structure their organization in many ways. There is a generally accepted group of functions which a successful firm must cover if the firm is to operate efficiently. Several functions may be performed by a single person, because of an individual's aptitude and interest.

In Figure 3.1, which follows, operation functions are segregated under the Association of Consulting Management Engineers (Acme) Handbook's outline. This publication is sometimes used by consultants in the analyzing of a corporation. Using this checklist (or a similar checklist) an analyst can determine fairly rapidly if all the functional bases are covered, and by whom.

Comparison of an existing corporate organization with a theorectical organization often shows the shifts of power within a firm and can tell of potential structural weaknesses.

SIMPLIFIES CHARTING

ORGANIZATION FUNCTIONS

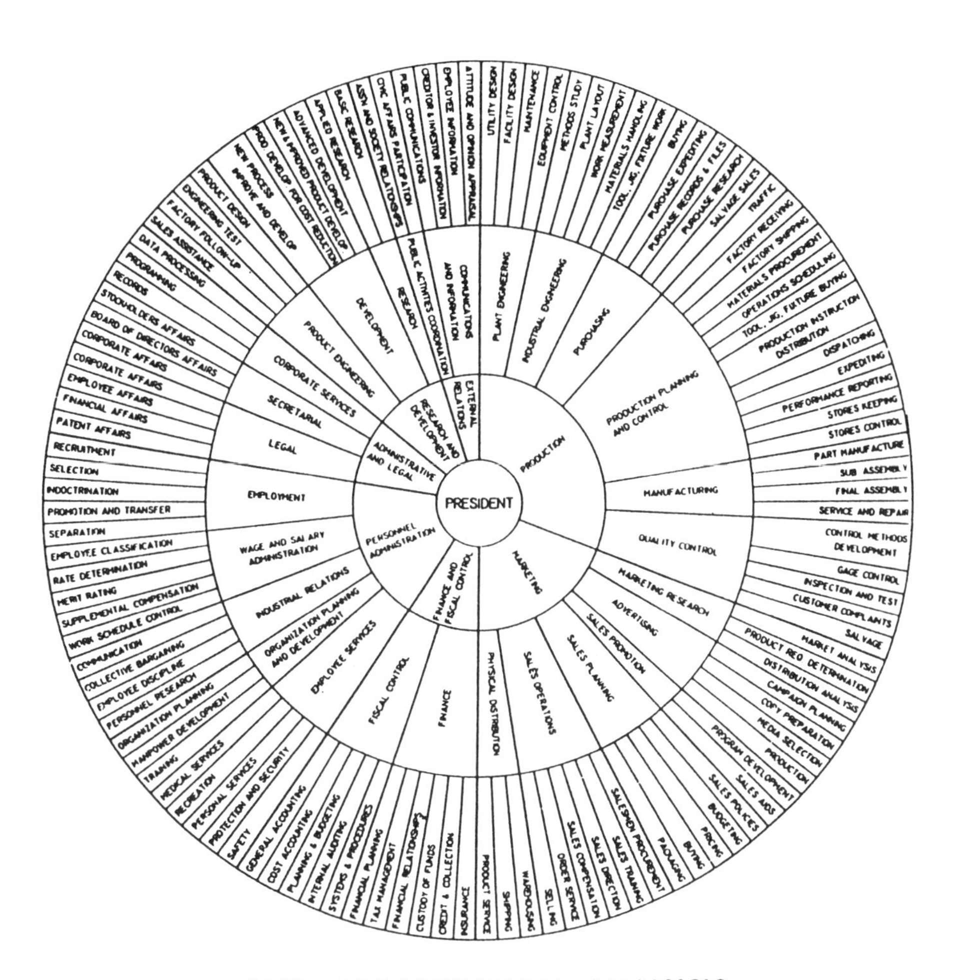

AND ORGANIZATION ANALYSIS

FIGURE 3.1.

The transmission of information to those who should know is an essential part of a smoothly run organization. Plotting the flow of data around the company will disclose omissions, excessive data dispersal and flaws in the company's operations.

ORGANIZATION FUNCTIONS

Figure 3.1 breaks the activity areas normally found in manufacturing organizations into its basic functions and sub-functions. The chart lists 117 commonly found functional divisions of activity. These functions and sub-functions have been placed under the activity area in which they are normally found. The chart could be further dramatized by coloring the Production activity area green, Marketing area red, Finance and Control blue, etc.

In reviewing a corporation's systems preparatory to developing new or advanced systems, functional relationships in respect to data flow, data use, and company orientation are necessary. As a supplement to this chart the ACME Report of September, 1957, (revised periodically) entitled "Common Body of Knowledge Required by Professional Management Consultants: Basic Organizations" is recommended. These charts describe, in detail, the functional responsibility or work content inherent within each activity area, function or sub-function. Use of this material facilitates organizational analysis.

In charting a firm you may find the Purchasing function (listed here as being a part of the Production area) under the Finance and Control activity area, or you may find Marketing Research reporting to Research and Development instead of marketing. In other cases, sub-functions such as Stores Keeping or Stores Control may be a part of the Control area under Finance and Fiscal Control. Color coding the sub-functions will rapidly point up the shifts of responsibilities and develop for you a pattern of the company with which you are working.

INFORMATION FUNCTIONS

Figure 3.2 deals with information flow on a conventional organization chart. It illustrates data flowing from some elements up to various users. Information must be examined as to point of origin, purpose, frequency of use, type of information, and number of users. Any given piece of information may have multiple use, either for the total segment of information, or for various portions of the information. As an example, the information as to the number of sales on hand can be broken down into units, total dollars, customers awaiting service and potential profits. It is conceivable that Sales would be interested in all of this information, that Production may be interested in the number of units to be produced and the customers to which they may be shipped, that Finance would be interested in the total dollars of business outstanding and the potential profit, that Administrative and Legal would be interested in the obligation to various customers, that Research and Development would be interested in the accumulations of units to be produced with a view to either

TO ANALYZE INFORMATION AS TO

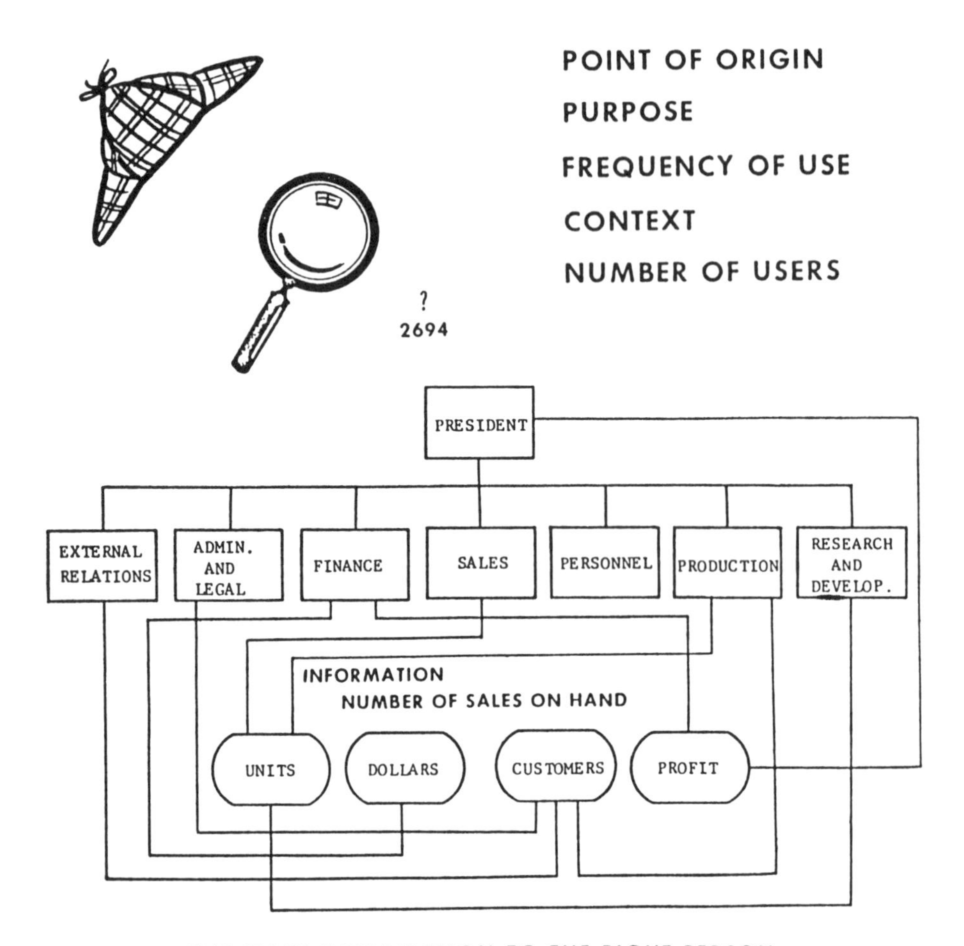

THE RIGHT INFORMATION TO THE RIGHT PERSON
FOR THE RIGHT FUNCTIONAL PURPOSE

FIGURE 3.2.

expanding or improving the product line or production line, that External Relations would be interested in potential publicity value of selective customers and that the President would be interested in the potential profit.

Selection of data so that the right function receives the information necessary to the proper execution of its job requires relating information to function. Before a new or advanced industrial management system can be built, capable of selecting the right information of the right person at the right time for the right functional purpose, a fair understanding of the purpose of the flow of information within a corporate structure is necessary. This may be obtained by charting reporting relationships and by following the movement of information. Reporting relationships will then be clarified. Systems design will proceed more rapidly as one learns where to go to obtain decisions or approval of procedure changes. Key people for successful installation of systems or for successful sales of ideas rapidly emerge.

TRACKING DATA MOVEMENT

Once the functions within the company have been defined and the reporting relationships classified, it becomes a less difficult matter to distinguish the necessity for data being used in certain functions. The functional chart (Fig. 3.3) can serve as a rapid means of tracking down, or flow charting, data movement.

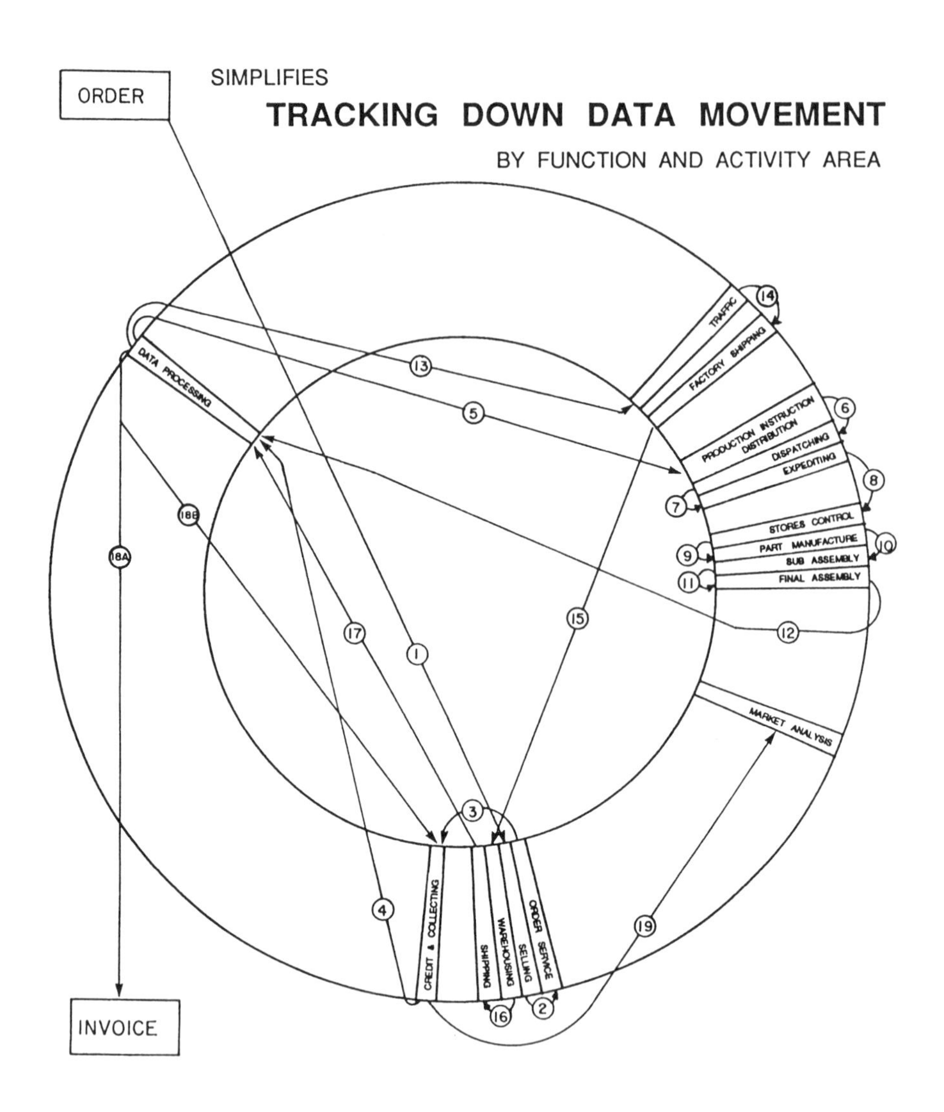
ORDER
SIMPLIFIES
TRACKING DOWN DATA MOVEMENT
BY FUNCTION AND ACTIVITY AREA
DATA PROCESSING
TRAFFIC
FACTORY SHIPPING
PRODUCTION INSTRUCTION DISTRIBUTION
DISPATCHING
EXPEDITING
STORES CONTROL
PART MANUFACTURE
SUB ASSEMBLY
FINAL ASSEMBLY
MARKET ANALYSIS
CREDIT & COLLECTING
SHIPPING
WAREHOUSING
SELLING
ORDER SERVICE
INVOICE

FIGURE 3.3.

In Figure 3.3 the movement of an order through Sales into Production through Production back into Sales and finally out as an invoice is illustrated. Nineteen steps are shown. The pattern depicted may or may not be typical for every order. Differences can be color coded to facilitate comparison.

Using a functional chart it is possible to diagram or flow chart data movement of a very complicated nature in minimum time. This flow charting provides a basis for analyzing data movement. The chart assists in determining what portions of information are important to what functions. Examination of the functional description of activity against the character of information embodied in the document charted provides a way to block out areas of importance in the form. Forms can then be simplified, combined or eliminated.

Analysis of information flows enables a substantial reduction of the paperwork within a corporation and substantial improvement in the quality of information in use. It also simplifies the job of placing systems on a computer. It must be kept in mind that the information needs of a company constantly change. The batch accounting systems (card-oriented) are no longer responsive to the demands for shortened production cycles and improved customer service. There are many computer systems using semi-on-line techniques which fail to protect information integrity and fail to provide timely information to functions needing action data.

FUNCTIONS OF CONTROL FOR PROFITS

Control is exercised through Systems and Procedures. Typical examples of anticipatory control procedures are: forecasting or planning; manpower, machine use, materials, money or budgets and purchases and evaluating activity against forecast (Fig. 3.4).

Typical of the controls for immediate reaction are: management reports on schedule progress, dispatching, receiving and shipping performance reporting. In each case a positive action must be taken when any deviations are reported.

Typical of the cyclical control functions are: machine loading, inventory maintenance, quality control checks and fiscal activities such as; payroll, monthly and quarterly financial reports, cost accumulation, etc.

The approaches to control systems combine:

a. Recordkeeping only, which is a form of historical control.

b. Supervision of operations, which is a form of manual or oral visual control.

c. Optimized operation which is a pre-determination of the best product mix to produce the most for the least cost as carried out through scheduling and dispatching activity.

DRAWS FROM TAILORED PROCEDURES

MANAGEMENT CONTROL DATA

SALES

SALES - PRODUCT SHIPPED

BACKLOG - PRODUCT ORDERED

FORECASTED SALES

SALES ANALYSIS

PRODUCTION

PRODUCT MADE

INVENTORY - RAW MATERIALS

WORK-IN-PROCESS

PARTS AND ASSEMBLIES

FINISHED GOODS

INVENTORY TURNOVER

FINANCE AND FISCAL CONTROL

CASH FLOW

PROFIT ANALYSIS - BY PRODUCT

BUDGET PREPARATION

BUDGET ANALYSIS

FIGURE 3.4.

Steps in exercising control are:

a. Maintaining adequate records.
b. Establishing control limits.
c. Developing action reports.
d. Following up instituted actions.
e. Utilizing exception reporting.
f. Preparing performance reports.

In today's widespread use of telecommunications, preserving data integrity and providing completion checks on staff entering, releasing and compiling data is becoming increasingly important. The wrong code can access the wrong computer and cause the wrong action. Of course, this is a different problem from computer data theft by industrial spies, teenagers, and others who can locate and break access codes.

ZONES OF CONTROL

The type and nature of records and information varies from sub-function to function on up to activity area in corporate management. In general, detailed day-to-day instructions, reports, activity measurements, schedules and analyses are required at the sub-function level. This data may be maintained on a PC or mini-computer (Fig. 3.5).

Daily detail is reviewed and filtered by foremen and managers before it is moved in summarized form to the function level. Computers may assist in data consolidation.

The process of review continues at the function level where semi-detail is summarized into area activity reports and

LOCATES

ZONES OF CONTROL

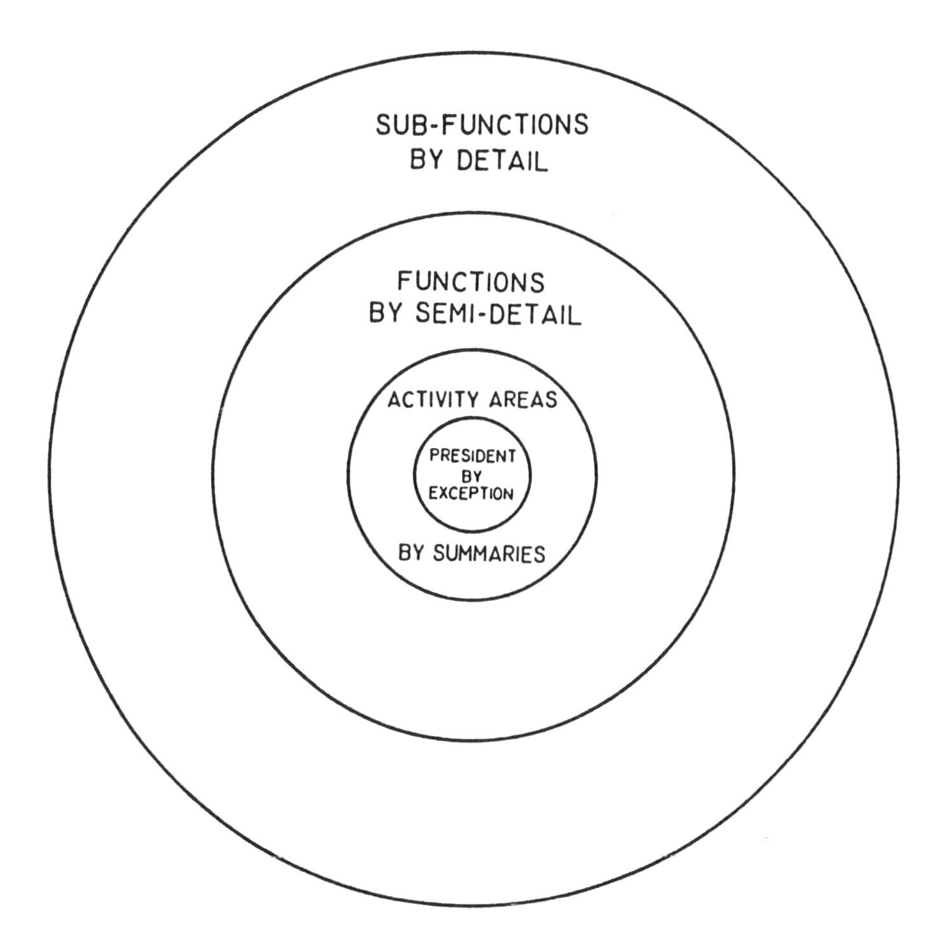

FOR IMPROVED SYSTEMS

FIGURE 3.5.

sent to the management of the activity area. Or a home office computer may poll a regional or specialist functional computer in order to bring up data by wire.

Activity area management again reviews and summarizes detail and prepares reports for top management. Usually this is done in the corporate headquarters. As information technology improves the reviews and summaries will be done by expert systems and the corporate president will have direct data access.

A general rule can be drawn which states: as functions are sub-divided, control detail increases, developing feedback systems for reporting becomes more complex. As reporting systems become more complex, increased supervision is required to keep the volume of detail appropriately filtered to useful management reports.

A computer system permits automatic summarization of detail through exercise of control limits and production of exception reports. This is a basic principle utilized in the procedures which will be reviewed in subsequent discussions.

MANAGEMENT CONTROL DATA

A good information program draws from tailored procedures to produce sound management control data. Latest installations of a new or an advanced industrial management system requires a distributive information approach. Tapes, floppies, or disc files can be used to accumulate corporate activity in such a fashion that management control reports can be quickly generated (Fig. 3.6).

As examples, typical sales reports may be: product shipped, back orders, forecasted sales versus actual sales and sales analyses by territory or geographic area.

Typical production reports may be: product made within specific time period versus product made in similar time periods in the past, inventory condition of raw material, work-in-process, parts and assemblies or finished goods and inventory turnover comparisons.

Typical finance or fiscal control reports may be: analysis of cash flow, profit analysis of the product line by product, budget preparation reports together with analysis of expenditures against budgets.

Management systems store tables, formulas, data in respect to trends and relationships of the constantly changing vital business factors. These systems will provide periodic management reports which will give management the opportunity to anticipate needs for changes in marketing, for expansion or contraction of working force, plant capacity, or changes in the manufacturing plant or process.

Traditionally, operating data is accumulated from transactions based on calendar periods related to financial activity. The daily maintenance of current operating records represents the opportunity to prepare major control

FUNCTIONS OF CONTROL FOR PROFITS

ANTICIPATORY

- FORECASTING
  - PLANNING
    - MANPOWER
    - MACHINE USE
    - MATERIALS
- BUDGETING
- OPERATION EVALUATION
- PURCHASING

IMMEDIATE REACTION

- MANAGEMENT REPORTING
  - SCHEDULING
  - DISPATCHING
  - PERFORMANCE REPORTING
  - RECEIVING
  - SHIPPING

CYCLICAL

- MACHINE LOADING
- INVENTORY MANAGEMENT
- MAINTENANCE
- QUALITY
- FISCAL

FIGURE 3.6.

reports as and when needed on a regular or exception basis.

This type of data enhances the management's ability to perform its task of supervision of corporate activity.

4

Corporate Goals

GENERAL

Boards of Directors in consultation with corporate management and various experts as deemed necessary are charged with the functions of establishing long- and short-term corporate goals. The results of their direction is in turn translated into company policy.

Generally speaking, strategic planning can take many forms. In any form accepted, systems, procedures and equipment must become responsive to the plan and must provide management with the control data necessary to implement the plan.

STRATEGIC VERSUS OPERATIONAL PLANNING

The benefits of sound business planning, both strategic and operational, are unmistakably clear. "Strategic" planning, however, must precede and be separated from short- and long-range "operational" planning.

Strategic planning is determining "what" an apparel organization wants to look like, or should be, in the future. Most short- and long-range "Operational planning" and decision making are determining "how to get there." The "what" and "how" dimensions of business planning should not be confused.

Strategic planning is the vehicle through which top management of an apparel company plots a course through competitive waters which appears most likely to ensure the future attainment of the volume, profit, and return on investment goals of the company. It is also a style of management. It is an organized process for providing explicit and effective answers to four basic questions: (See Fig. 4.1 - The Strategic Planning Process.)

a. Where are we now, as a going business?

   Growth, static, decline, competitiveness, strength, weakness.

b. What have we available?

   Resources, products, customers, consumers, facilities, people, image, momentum.

c. What do we want to be, and where do we want to go?

   Objectives, goals, growth, profitability, directions.

d. How do we get there? (Strategic and Operational)

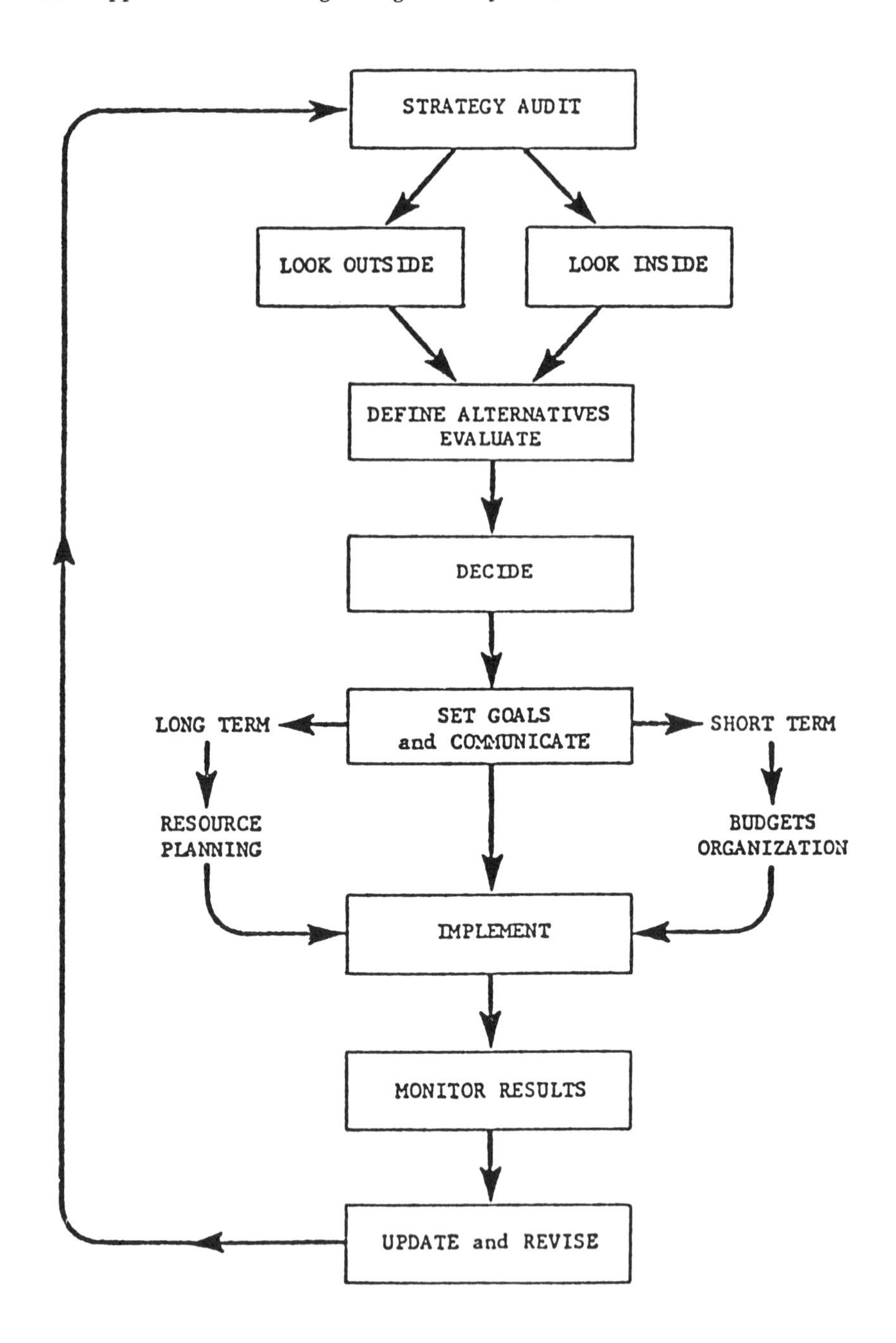

FIGURE 4.1. THE STRATEGIC PLANNING PROCESS

Strategies, plans, tactics, resources, timetables, policies.

"Strategy" is a framework that guides those choices determining the nature and direction of an apparel organization; thus, strategic planning is a function of direction, not time.

"Strategic" areas generally can be grouped in the following categories:

a. Products or services, geographic markets, and customer groups.

b. Capabilities of technology, production, and methods of sale and distribution.

c. Results in size, growth, profit, and return on investment.

It is only by separating "strategic" from "operational" thinking that management can intelligently assess which products, services, and markets should be emphasized or abandoned, and determine what the scope of new products, markets, or services should be.

Put simply, the purpose of strategic planning is to create competitive advantages. Good strategic planning looks at competition as a "system" and seeks to "influence the market" by creating competitive advantages rather than being left to the mercy of the market.

A foundation for good strategic planning lies in what can be termed a firm's "mission statement." A "mission statement" expresses a firm's internal objectives which relate to survival, profitability, growth, and external objectives such as market share, market segmentation, or attracting and keeping a profitable customer. What ties these objectives together in a "mission statement" is keeping in mind the "desired" competitive position. (See examples of "mission," "philosophy," "charter," etc. in the following sections.)

Once strategic objectives are established, top management's task is the allocation of resources, including capital, capacity, and personnel, to achieve those objectives.

In large corporations, resources are allocated among business segments called strategic business units (SBU). In apparel, a SBU is a marketing concept related to a common customer group and/or product line which requires a separate strategy. Within each SBU, depending on its structure, multiple or various profit centers can exist for financial reporting, management control, or other purposes.

Management must decide on which investment strategy to use for each SBU:

a. Invest aggressively to make the unit grow aggressively?

b. Invest selectively to enable the unit to grow at a moderate rate?

c. Invest selectively to maximize earnings, but not generate growth?

d. Remove resources from the unit, thereby "shrinking" or "harvesting" it?

To make such decisions, top management must understand each strategic business unit's competitive position, markets, and performance potential. Management then must prioritize the allocation of resources among SBU's to accomplish the most effective utilization of the corporation's available resources to achieve overall corporation objectives.

An example of examining and evaluating performance potential in a typical company division to determine the allocation of resources among strategic business units (SBU's) is outlined in Figure 4.2. This figure shows goals and historical performance.

Sound short- and long-range operational planning, or the "how to get there" dimension, is vital to an apparel organization in achieving its strategic objectives. ("Operations Planning" and "Establishing Objectives" are discussed in a later segment of this course.)

MISSION STATEMENT (Example)

ABC Industries, Inc.
The Mission

The management group of ABC Industries, Inc., has been assigned by the corporation the responsibility and accountability for the marketing, manufacturing, and distribution of the company's products. It is the mission of

SUMMARY OF SELECTED FINANCIAL DATA & TRENDS

(In Millions of Dollars)

	Historical Performance					Expected Performance	
	1978	1979	1980	1981	1982	1983	1984
NET SALES							
Amount - Sears	$ 30.4	$ 34.5	$ 34.9	$ 47.4	$ 56.5	$ 55.0	$ 61.3
- Trade	0.6	0.5	0.8	0.2	-	-	-
- Total	$ 31.0	$ 35.0	$ 35.7	$ 47.6	$ 56.5	$ 55.0	$ 61.3
Pct. Increase - Total	#	12.9#	1.0 #	33.3 #	18.7 #	< 2.7 > #	11.5 #
PROFITABILITY							
Operating Income	$ 2.1	$ 2.9	$ 4.0	$ 3.9	$ 5.6	$ 4.9	$ 5.4
- # Sales	6.8 #	8.3 #	11.2 #	8.2 #	9.9 #	8.9 #	$ 8.8 #
Earnings Before Tax	$ 1.5	$ 1.9	$ 2.4	$ 2.4	$ 3.5	$ 3.1	$ 3.5
- # Sales	4.8 #	5.4 #	6.7 #	5.0 #	6.2 #	5.6 #	5.7 #
INVESTMENT & RETURNS							
Total Avg. Net Assets	$ 9.0	$ 10.2	$ 13.3	$ 15.5	$ 17.7	$ 18.5	$ 20.1
-Turnover	3.4 x	3.4 x	2.7 x	3.1 x	3.3 x	2.9 x	3.0 x
Return on Avg. Net Assets	16.7 #	18.5 #	18.1 #	15.4 #	20.4 #	16.3 #	17.3 #
< Inc > / Dec Loan Balance							
- Annual Amount	$ 1.7	$ < 0.4 >	$ < 1.5 >	$ 1.7	$ 0.9	$ 2.8	$ 0.7
- Cummulative		1.3	< 0.2 >	1.5	2.4	5.2	5.9
SELECTED ANNUAL RANKINGS							
Net Sales	5	5	5	2	1	2	2
Earnings Before Tax	5	1	2	3	2	2	3
- # Sales	14	5	7	10	7	7	7
Total Avg. Net Assets	7	7	5	3	2	2	2
Return on Avg. Net Assets	7	5	7	10	6	8	9
* Total SPU's each year	27	29	29	28	27	26	26

AVERAGE GROWTH & PROFITABILITY

	Historical		Estimated
	All Years	Recent Two	Next Two
• Growth in Sales	16.2 #	25.8 #	4.2 #
• Percent of Sales			
- Operating Income	9.0	9.1	8.9
- Earnings Before Tax	5.7	5.7	5.7
• Return on Avg. Net Assets	17.9	18.0	17.1

* Forecasted results for 1982

FIGURE 4.2.

the company to offer products of such value to the user as to entitle the company to expanding volume and profits. We will serve the changing wants and needs of our customers as faithfully as we can in order to obtain an expanding share of the market.

Our corporate sales growth objectives will be established on an annual basis and reviewed or revised semi-annually, recognizing current and near-term economic conditions, as well as apparel market and product sector trends and activity. Our long-term pretax profit goal will be no less than 5 percent of average assets in use. Short-term profit objectives will be determined as an integral part of our semi-annual operations planning and budgeting function.

Various operating departments, functions, or management centers may develop objectives from time to time that differ from overall corporate objectives or priorities. Corporate funds will be prioritized and allocated to those objectives yielding the best return on investment. Within the framework of the corporation's ability to fund specific program objectives, resources will be allocated and reallocated among operating departments, functions, or management center to take advantage of unique opportunities to increase sales or reduce operating costs.

While we must maintain acceptable levels of sales and profits on an overall basis to ensure financial stability and to satisfy the corporation's stockholders, actions which maximize short-range sales and profit performance will not be taken at the expense of the long-range health of our business.

We will recognize that the ultimate measure of a business enterprise is the economic result it produces.

CORPORATE PHILOSOPHY (Example)

ABC Industries, Inc.
The Philosophy

We will maintain high ethical standards in external and internal relationships. When there is any doubt about which action to take, we will rely on the guidance of ethical principles. Actions which are unprincipled will not be condoned. All laws, government regulations, and company policies will be observed in the conduct of our business. All decisions will be based on facts, objectively considered. We will concentrate on "what is right," not "who is right."

We will be outward looking and sensitive to the market and the external forces in our environment, especially as they relate to the ever-changing needs, attitudes, and direction of our customers on whom our success ultimately rests.

We will administer our business with a sense of competitive urgency, facing problems as they arise, and seizing new opportunities as they become available. We will become neither complacent nor arrogant in the face of success, nor will we accept mediocrity in our commitment to improved performance. People are our most valuable asset, and the management of the corporation has a binding commitment to encourage the pursuit of excellence by all employees and to uphold their human dignity. Within the limits of prudent business judgment, we will strive constantly to maintain level and stable employment for our employees. People will be judged on the basis of their performance, not on personality, education, or personal traits.

Wherever possible, we will promote from within the company and will encourage transfer of promotable employees between operating departments, functions, or management centers when equivalent opportunities are not available within the employee's department or functional management center.

The various functional elements of the corporation will work together harmoniously with complete exchange of information and facts and will be dedicated to the corporation's common objectives, goals, policies, and plans.

DIVISIONAL MISSION STATEMENTS (Examples)

Many large firms are organized by division or product group. Readers of THE WALL STREET JOURNAL, BUSINESS WEEK, and other business publications find corporations constantly regrouping their products and product activity. The obvious reason to regroup is to strengthen the firm - either by reducing overhead or by strengthening an area of promise while deemphasizing an area showing weakness. Two typical division statements follow for one division of ABC Industries, Inc.

Menswear Group
Slacks and Jeans Division
Marketing
Charter

Mission: To create and market fashionably styled men's apparel, of recognizable quality and value, timely in market impact, manufactured and delivered on a consistently superior basis, and priced competitively within the framework of the financial objectives of the company.

Product Assignment: Men's Apparel

Primary: Men's casual slacks (both woven and knit); men's basic and fashion jeans.

Secondary: Shorts

Opportunistic: Exploration of other men's products as market validity is determined involving potential volume, risk, and profitability.

Market Assignment: National major chains, national/regional speciality chains, national major discounters, promotional department store groups, and selected special accounts which purchase on a contractual or firm order basis.

Method of Selling: Salaried sales executives and commissioned sales representatives.

Production Sources: Assigned corporate domestic manufacturing facilities, outside domestic contractors as required, and offshore procurement (if desirable) on a finished product basis.

Menswear Group
Slacks and Jeans Division
Manufacturing
Charter

In its function, the Slacks and Jeans Division's Manufacturing operation of ABC Industries, Inc.'s Menswear Group shall subscribe to the statement of mission and philosophy of the corporation, of which it is a part.

The function of the manufacturing operation is to manufacture products of value to meet the marketing needs of the company. The products shall be of value to our customers and the ultimate consumer insofar as they are in the span of control of manufacturing, in that:

Quality: They will represent consistently superior quality.

Service: They will be entered and processed on a timely basis.

Cost: They will be produced efficiently at a competitive cost.

The manufacturing responsibility, as defined by corporate policy, begins with the receipt of piece goods and ends with the delivery of finished products to the company's trucking fleet for transportation to division distribution centers.

It encompasses the responsibility for the quality of all product components, including fabric and findings, and of the finished products in their conformance to specifications, package appearance, and seaming characteristics.

It includes the mutual responsibility, with the merchandising and styling function, for the successful implementation of new production patterns, considering style and design objectives, specifications for fit, sewability, and product appearance.

The manufacturing function involves the responsibility for the time required in the manufacturing cycle and for the timely and accurate estimation of completion dates for production work orders.

It involves the control of costs, of labor and overhead, and of waste, including material, findings, and manufacturing irregulars. It includes the effective and efficient scheduling of the facilities, machinery, and equipment, the recommendation of the purchase of equipment in the pursuit of progressive manufacturing techniques to lower costs, and the timely and accurate costing of the manufacturing process for product costing.

The manufacturing operation's responsibility includes the warehousing of material and findings and the maintenance and reporting of all necessary records and operating controls.

It encompasses planning and the implementation of the manufacturing operation's segment of the Slacks and Jeans Division operations plan; it encompasses the setting of objectives which will ensure the continuing manufacturing competitiveness of the division in the market.

To implement its function successfully, the manufacturing operation will exercise appropriate responsibility to all its employees in the administration of good personnel practices and procedures and will provide for the growth and development of its human resources toward the fulfillment of potential.

Finally, the manufacturing operation shall perform its function in the best interests of the corporation and in cooperation with all other divisions and/or corporate functional operations in ABC Industries.

OPERATIONAL PLANNING AND ESTABLISHING OBJECTIVES

General

We have examined briefly the difference between strategic and operational planning, described "what" strategic planning is, and why it must precede and be separated from short- and long-range operational planning. (One, Strategic Planning, involves Strategy and Direction, and the other, Operational Planning, involves Tactics and Time.) How do they relate? An example follows.

Organizational Practice in Strategic Goal Setting (Example)

The sequence of specific events that composes strategic goal setting can be pictured in the form of a calendar. These are the events that must happen this year if next year and succeeding years are to be better, or if we are to avoid the consequences of changes that could threaten us. The three key dates for strategic goal setting are pictured in the exhibit in Figure 4.3.

1) Audits that are ordinarily completed about the end of the third month of the year are the first take-off point for strategic goal setting.

2) The annual Business Plan or "Annual Edition" of the five-year strategic plan, will be needed by midyear for the coming year (starting six months hence) if all of the necessary staff work and decision making are to be in place.

3) Annual budgets should be submitted for higher management review and approval by the early part of October so that decisions may be made and appropriations in place for the start of the following year on January 1.

This three-year sequence is the major element of the strategic goal setting cycle. Of course, numerous other things need doing as the year goes along, and many operating decisions and actions must occur. Our purpose here is simply to outline the strategic goal setting process and how it relates to operational planning and goals. (See Figure 4.3.)

ANNUAL PLANS - A PLANNING SCHEDULE EXAMPLE

The typical "Annual Plan" process is timed to permit the integration of strategic with operations planning. AAMA has issued several pamphlets on this process. Following is an example for fiscal 1983.

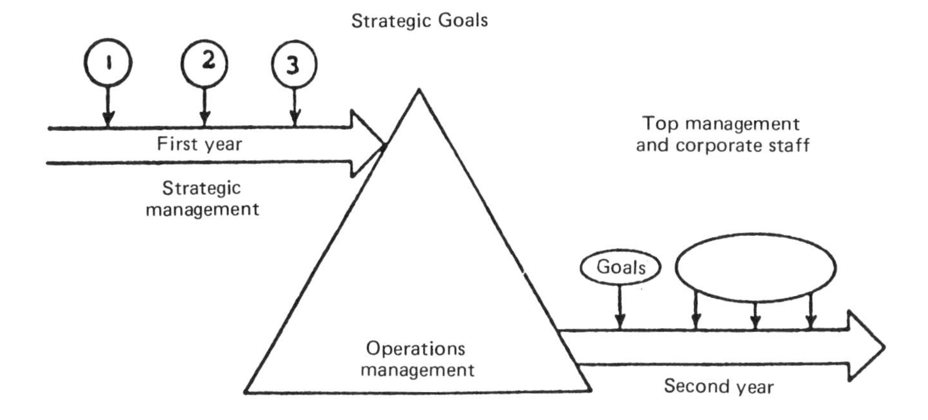

FIGURE 4.3. STRATEGIC GOALS AS RELATED TO OPERATIONAL GOALS

October 1981 - Strategic Planning Conference (Groups/Units present their Strategic Planning Objectives)

November 1981/January 1982 - Group/Functional Annual Planning

February 1, 1982 - Annual Plans to Headquarters

March 1-15, 1982 - Group Annual Plan Review

April 15, 1982 - Final Approval

April 23, 1982 - Annual Plan Presentation to the Board of Directors

May 1, 1982 - Beginning of Fiscal Year 1983

Annual plans are designed to focus on each operating group's total business and on their individual business units. The annual plan is developed primarily for the benefit of each group and to assist each group in the conduct of its businesses.

The role of corporate top management is to review and analyze these plans and to consolidate them for purposes of resources allocation, financial funding, performance evaluation and measurement, and assistance and guidance toward corporate goals. Each group, while adhering to corporate planning department guidelines to ensure uniformity, is essential to corporate planning requirements, exercises flexibility and initiative in development of annual plans that recognize their unique circumstances, conditions, and objectives.

Thus, the corporate annual plan becomes an integral part of the strategic planning process each year and is a vital step in the company's continual planning process.

PLANNING AND THE DYNAMICS OF "CHANGE"

Economic conditions, demographics, technology, human resources, markets and products, financial strategies, strategic and operational planning -- the list of apparel industry challenges could be expanded almost endlessly.

What is the common thread which weaves itself through each challenge? What common denominator binds each of these multi-faceted challenges to each other? The Dynamics of Change! The ability to make or become different. The ability to effect a different position, course, or direction. Not change for change's sake, but that which transforms, improves, and provides lasting benefits for the business enterprise.

A willingness and desire to accept, indeed, to welcome creative, constructive change at all levels of an apparel organization are vital to its future growth and success.

It is management's responsibility to "manage change" within the framework of reality, in an orderly manner, exercising sound judgment, and to relate the "dynamics of change" to the objectives of the organization. It is management's unique imperative to create, foster, encourage, promote, and establish an environment in which the spirit of dynamic, constructive change is welcomed, understood, and capitalized on to its fullest advantage.

The resulting benefits, measured in both human terms and the future growth and success of the business enterprise, will be rewarding and long lasting.

OPERATIONS PLANNING - GENERAL

a. All business planning begins with definition.

b. It is goal, or objective, oriented and states them in clear, concise terms.

c. It outlines the people, methods, means, resources and time frames required to achieve the objectives.

d. It is not confined to large, monolithic corporations -- but is an activity applicable to the smallest of businesses.

e. It encompasses every facet of a business enterprise.

f. It expresses both short- and long-term goals.

g. Its goals are both financial and nonfinancial.

h. It is adaptive and flexible in its approach and implementation.

i. It recognizes the dynamic nature of change.

j. It is not a "security blanket" to be followed dogmatically.

k. It represents a moment in time, a frame of reference, a point of departure.

l. It is a management tool, utilizing sound judgment, to maximize return on investment, to ensure the sound application of resources, and to capitalize on opportunities for future growth in existing or new products and/or markets.

OPERATIONS PLANNING - ESTABLISHING OBJECTIVES

One of the greatest deterrents to successful operations planning is the feeling on the part of many apparel managers that it is "too restrictive," or "poured in concrete," or that conditions and circumstances are changing too rapidly for reasonably accurate projections and planning.

A sound operations plan is not a security blanket to be followed dogmatically. It is a management tool that represents a moment in time, a frame of reference, a point of departure, and is subject to adaptive, flexible change before the ink is dry. Just as a plane or ship changes speed, course, and direction when encountering unexpected adverse weather conditions, a good operations plan is quickly modified with adaptive planning and strategies.

Adaptive planning requires that we be prepared to modify our objectives in a changing environment. That statement of strategy would tend to persuade us that an objective has a somewhat temporary "shelf life." It must be temporary if we agree that we cannot really predict the future with accuracy.

Webster defines an objective as "something toward which effort is directed; goal, or object." Future objectives can be accomplished only if adequate resources--people, money, material, time, and know-how--are provided in the right quantity and quality, and in the right time and place. An objective without essential resources is not an objective--it is a delusion.

An amplication of the definition of "objectives" and an identification of the criteria which should be used to evaluate "an objective" will help us to determine whether our objectives would be effective or not.

OBJECTIVES

An "objective" is a temporary estimate of a desirable future result that cannot be predicted with accuracy, that you are willing and able to pay for, and that you believe you can achieve through your effort.

Criteria for Effective Objectives

Objectives must be suitable: This is of primary importance. Is the objective suitable? If it were achieved, would it take you in the right direction? Is it relevant, and does it support the purpose and mission of the organization? Is it compatible with other objectives to which you are already committed? Can it be coordinated with other objectives to support the objectives of the next higher echelon? Unless the answers to these questions are affirmative, your objectives should be either modified or discarded to avoid frustration or the squandering of resources or time.

As an example of this, suppose the owner of a country store has the objective of increasing his sales by 20 percent next year. His clerk suggests that he buy the gas station next door. This purchase might enable him to get his 20 percent sales increase, but if he does not want to get into the gasoline business, it is not a suitable objective--because it does not take him in the direction he wants to go.

Special care and attention should always be given to any objective or program that tends to change the nature of a business.

The objective must be feasible: Is it possible? Can it be attained? While there is much to be said in behalf of courage, enthusiasm, and the old college try, such attributes can cause more harm than good if they permit commendable emotions to override common sense or encourage the commitment of time, effort, and money to objectives that are actually unattainable--regardless of how desirable they may be.

The establishment of objectives involves decision making and the quality of decisions cannot consistently rise above the quality of the information or knowledge on which the decisions are based.

You cannot afford to have all the information you would like, but you ought to get the best information you can afford--and you need to get as much as necessary to make the decision with which you are confronted and make it with acceptable risk.

The objective must be acceptable: Are you willing to pay the cost of this objective? Or, more specifically, are the "owners that be" willing and able to provide the essential resources--in the right quality and quantity, at the right time and place, and at the right cost? More planning and "objective setting" is off track on this score than on any other.

The objective must be valuable: Is the objective worth the price?--and is it the best buy you can get for your money? Resources must be used for maximum effectiveness, and no organization has enough resources to support every objective that is suggested. That is why priorities are established that balance effectiveness, needs, and resources.

The objective must be achievable: Is it achievable by you? How? This criterion is substantially different from the second item regarding an objective's feasibility. It is one thing for an objective to be theoretically possible, but quite another matter for you to be able to achieve it yourself. For example, any high school athlete easily can jump 5 feet over a crossbar, but it may be difficult for his father, and impossible for his grandfather.

Results are often disappointing because of failure to discriminate between those objectives that may be feasible for someone else--and those that are achievable by you. Also important is the determination of how an objective will be achieved before approval is granted and resources are allocated. Here is where objective setting and planning truly interrelate. It is easy to set an objective to increase sales by 15 percent. The planning begins when we determine how we will achieve this goal--who sells what, to whom, when?

An Objective Should Be Measurable: Is it conveniently possible to measure progress against the objective? Can it be quantified in terms of quality, quantity, time or cost? Is it possible to reach agreement on the method of measurement so the responsible person can exercise self-audit and minimize the need for externally imposed control?

An Objective Should Be Adaptable: Is it possible to modify the objectives in response to unforeseen developments so that performance can be optimized in a changing environment? This criterion has become increasingly important because of the rapid increase in the rate, type, and volume of change in business environments.

An Objective Should be Capable of Inspiring Commitment: Whether objectives are "proposed" from the bottom, or "imposed" from the top, once agreement is reached, there must be a firm commitment on both sides to provide the resources and to accomplish the result. The first five criteria for objectives are essential--the last three are highly desirable.

An objective is not truly an objective, and it should be neither proposed nor approved unless it meets the first five standards. But, if unexpected developments occur, objectives may no longer meet these criteria. When that happens, the last criterion should be cancelled.

It is unwise to continue a firm commitment to an objective that is no longer suitable, feasible, acceptable, valuable or achievable.

Summarizing Objectives

An objective is a temporary estimate of a desirable future result that cannot be predicted with accuracy, that you are willing and able to pay for, that you believe you can achieve through your effort.

Objectives must be		
	suitable	Essential
	feasible	
	acceptable	
	valuable	
	achievable	

Objectives should be		
	measurable	Desirable
	adaptable	
	capable of inspiring commitment	

5

Corporate Marketing Strategy

INTRODUCTION

Corporate goals include both long- and short-term market and product strategies. As is mentioned under the organization chapter these strategies often result in corporate reorganization, investment and disinvestment in properties, products or plants.

In many firms changes in market or product strategies are held confidential until execution of the change is reasonably well underway. If results do not appear to be positive, firms may reverse a strategy and entertain a new approach.

STRATEGY AUDITS

Managements of marketing and manufacturing firms normally conduct periodic strategy audits or strategic planning through diagnostic analysis. This is done in an effort to compete as effectively and profitably as possible. Decisions are made which affect the future course or composition of their business.

Management usually finds that there are four basic strategic product and market options available to them, within which all "sub-strategies" are facilitative:

a. The same products, sold in the same markets.

b. The same products, sold in new markets.

c. New products, sold in the same markets.

d. New products, sold in new markets.

One, several, or all of these basic strategies can be employed as an outgrowth of strategic product and market planning. Base sales data describing a total market can be obtained from a computer service firm.

Creative, innovative, as well as basic, practical, and realistic alternatives and options for modification, deletion, addition and/or diversification of a company's product mix and market target sectors are identified, explored, and evaluated for management consideration in developing market and product strategies.

For example, adaptive marketing strategies frequently can direct marketing efforts in identical, similar, or related products to different user/customer and/or geographical markets.

To identify new product opportunities (for current or new markets) a company's current product mix must be examined for relevance and effectiveness. New product opportunities also should be evaluated for, but not limited to, the following:

a. Volume potential.

b. Contribution to company's gross margin objectives.

c. Seasonality (inventory management challenges).

d. Financial risk (start-up vs. sales "life").

e. Relevance to current and future market activity.

f. "Positioning" relative to import and domestic competition.

g. Manufacturing capabilities/restraints.

MARKETING RESPONSE

Key elements of a sound product/market strategy response include a clear strategic plan based on:

a. Market segmentation and analysis.

b. Brand or trade marketing programs (if applicable)--focused consumer targets.

c. Product line management--focused line development.

d. Market organization development--sales and advertising.

Dissecting the marketplace as outlined above enables apparel firms to gain necessary understanding of the market, its trends, their own positions in it, their competitors' position in it, and finally, the opportunities the market offers.

This process of market definition, segmentation, and positioning is normally part of any continuously successful effort to develop a meaningful product/market strategy.

Management must recognize, identify, and quantify, however, the risk, time, resources, skills and assets required to implement viable marketing approaches or product/market alternatives and opportunities.

COMPETITIVE "SOURCING" (MANUFACTURING OPTIONS)

As an integral part of strategic and operational planning, U. S. marketing and manufacturing firms increasingly must examine all manufacturing options or alternatives open to them in developing manufacturing strategies.

Changing labor and raw materials sources, dictated by world economics and policies and combined with rapid technological changes, indicate that "sourcing" strategy will be a critical success factor in the 80's and essential to most major industry/product sectors (including the apparel industry).

The purpose of evaluating all sourcing options is to improve profitability and to optimize a firm's responsiveness to market opportunities and its competition.

"Variable" sourcing, however, presents major trade-off options between cost and flexibility. Some of the major sourcing options an apparel firm must examine follows.

U. S. Production of Apparel

Domestic Sourcing Only: Internal capacity and/or domestic contractor)

a. Advantages.

b. Disadvantages.

c. Competitiveness at current level of engineering and technology.

d. Potential with application of high-level engineering and technology.

U. S. and Foreign Sourcing:

a. U. S. Production and 807 (with and without equity).

b. U. S. Production and foreign supply (with and without equity).

c. U. S. Production, 807, and foreign supply (with and without equity).

All Foreign Supply:

a. U. S. design, merchandising, marketing, and distribution (all garments manufactured 807 and/or wholly foreign made).

"INTERNAL" SOURCING

Key elements of a proper sourcing strategy response for U. S. companies with their own (U. S.) manufacturing capacity include:

a. Technology upgrades.

b. Manufacturing controls of input, output, quality.

c. Productivity improvement/cost reduction.

d. Material usage controls.

e. Full human resource development.

"EXTERNAL" SOURCING

The degree of potential application of "external" sourcing alternatives, including U. S. domestic contractors, 807 manufacturing (with or without equity), or foreign supply (with or without equity), as well as "all foreign supply" sourcing, varies widely from one firm to another based on the firm's markets, products, price points, resources, and strategic objectives.

Protection of price points no longer achievable domestically, "promotional" items, new or complementary products, unique fabrications, styling, features, or lower cost/increased markup (improved profitability) are but a few of the reasons for the consideration of "foreign sourcing" options.

In examining and evaluating their sourcing strategy, U. S. apparel firms should:

a. Develop actual and future potential costs and investments for domestic and foreign sources (products and services).

Recent comprehensive studies (1982-1984) in the men's shirt industry and in the men's, boys', women's, girls', and infants' sweater industry (to establish the current and future potential competitiveness of these industries versus Far East imports) indicate its competitiveness versus Far East imports.

The studies show that this competitive edge can be achieved via U. S. investment in plant engineering and technology, while the Far East has only limited options.

While the application of engineering and technology yields a comparable potential productivity "percentage" increase in both the U. S. and the Far East, this percentage applies only to labor-related manufacturing costs,--thus reducing U. S. manufacturing costs (where labor amounts to a much higher percentage of the manufacturing cost than in the Far East) more significantly than the landed cost effect of similar manufacturing reduction percentages in the Far East.

Key factors to determining that these two apparel product sector industries have major future potential to become competitive with imports are:

a. U. S. investment in engineering and technology can result in significant cost reduction.

b. Far East costs have risen faster than U. S. costs during the past five years.

c. Future cost projections indicate that Far East costs will rise 2.5 percent faster than the U. S. during the next ten years.

d. The "Package of Services" a U. S. manufacturer can offer versus imports offsets a significant portion of the import cost advantagc.

e. Retailers incur a higher percentage of off-price sales with imports than they do with domestic products.

The above points were drawn from a case study involving the U. S. sweater industry. It is essential to develop actual and future potential costs as well as the investments for domestic and foreign sources as a part of establishing a competititve sourcing strategy.

SOURCING SUMMARY

What is the significance and relevance of "competitive sourcing," or the strategy of examining all "manufacturing options" to an apparel plant manager, manufacturing vice president or to a company Board of Directors?

To repeat:

a. "In examining and evaluating their sourcing strategy, U. S. apparel firms should:

   (1) Develop actual (current) and future potential costs and investments for domestic and foreign sources (of production)."

b. "Internal" (U. S.) sourcing. "Key elements of a proper sourcing strategy response for U. S. companies with their own (U. S.) manufacturing capacity include:"

   (1) Technology upgrades.

   (2) Manufacturing controls of input, output, quality.

   (3) Productivity improvement/ cost reduction.

   (4) Material usage controls.

   (5) Full human resources development.

Corporate overhead management is strongly involved (whether directly or indirectly) with each of the above key elements.

A U. S. apparel company cannot intelligently assess its manufacturing options unless it can: (a) accurately determine its current manufacturing costs, (b) develop future "potential" costs for U.S. production of its products, and

(c) relate these costs to other manufacturing alternatives. This assessment activity requires accurate basic costing systems now normally found on in-plant computers as well as on "home office" central computers. The degree to which a U.S. apparel company can reduce its U.S. manufacturing and overhead costs and can improve productivity in the future will vitally affect the comparison of its U. S. operating costs with other alternatives.

During the 1980s, companies in many industries will make major decisions affecting their future course and direction based on "competitive sourcing" evaluations. The ultimate success of U. S. apparel firms is determined by how effectively manufacturing management uses the major resources committed to an apparel manufacturing facility -- people, equipment, buildings, land. How effectively these resources are used to achieve improved productivity is vital to the future growth and success of the apparel/textile industries.

MARKETING AND SOURCING INTEGRATION

Merchandise Control and Distribution

The integration of marketing and "sourcing" is complex and critical (particularly if a firm utilizes multiple sourcing), and it involves high working capital risks. Inventories represent the single largest investment in apparel firms, frequently 50 percent or more of total "net worth." Computer control over multiple source activities is critical to cost and inventory control. Satellite transmission of data is increasing in multi-national firms.

Without effective merchandise control, an apparel business will not succeed. What is gained through successful market/product strategies and competitive sourcing can easily be lost through poor merchandise planning and control.

Most apparel firms operate in an environment of high stock keeping units, many customers, variable demand, and seasonal products of limited fashion cycle "life." They must deliver quality products on a consistently timely basis and achieve high working capital turns, high utilization of their sourcing (manufacturing) alternatives, and minimal inventory write-downs.

Key elements of a sound merchandise control and distribution "response" to accomplish these goals include the following information and control strategies for:

a. Product line planning, development, management forecasting, and inventory management "targets" to execute product/market strategy.

b. Material requirements planning, master scheduling, work-in-process controls, and forecast controls necessary to execute sourcing (manufacturing) strategy.

c. Order acceptance, processing, and control; release to ship, inventory accuracy, good cost and financial accounting; and performance measurement of control systems necessary to execute financial control.

Proper merchandise control integrates marketing and manufacturing so that an apparel firm's goals for delivery and facility utilization are achieved with a minimum investment in inventory.

An effective merchandising control system impacts all areas of an apparel marketing and manufacturing firm. The elements of this system must function on an integrated basis and be structured to be highly responsive to changing input/output requirements for management control, and/or intervention and decision making.

Major elements of a merchandise control system are in:

a. Merchandising/line planning.

b. Inventory management targets.

c. Forecasting and forecast control.

d. Master scheduling and controls.

e. Distribution control.

f. Organization.

The key elements of merchandising/line planning, inventory management targets, forecasting and forecast control, and master scheduling and control will each be covered separately in the segments that follow.

6

Preseason Planning (Merchandising/Line Planning)

INTRODUCTION

AAMA over the past several years (1980-1985) has repeatedly prepared information covering the importance of preseason planning to an apparel firm's success. Regardless of product source, a seasonal plan must be conceived and executed by firms within time constraints which are becoming increasingly tight.

The AAMA Systems Division prepared a review of the part computers can play in executing a "timely response" plan. (1984 Bobbin Show Presentation).

A fashion firm with six or more seasonal product lines to conceive, buy, make, and deliver to multi-retailer locations is faced with a very complex coordination problem.

With few exceptions, the vast majority of apparel manufacturers in the United States today are involved to some degree in styling, merchandising, and producing "fashion" apparel.

Their experience over the past decade or two has shown that American consumers insist upon exercising greater freedom of choice in the way they dress. Traditional dress codes have given way to the expression of individuality.

These developments have been good for the apparel industry overall. Under the stimulus of constantly changing and expanding life styles, new leisure time activities, and new uses for apparel, more "fashion" and style apparel is being made and sold.

All this has brought a significant increase in style variation to apparel. Fewer and fewer companies are engaged today in manufacturing a limited line of staple apparel products. Whether firms like it or not, most apparel companies are finding themselves in the "fashion" business, or at least dealing with more style variations. This trend will continue, bringing with it more challenges and more opportunities for every apparel marketing and manufacturing company.

These challenges and opportunities must be met with the development of a greater understanding and cooperation between the merchandising, sales and manufacturing functions of an apparel company.

An essential first step in determining what needs to be accomplished to strengthen the merchandise planning and controls function is to improve the understanding, cooperation and interfaces between the merchandising and manufacturing functions. This requires a diagnostic analysis, or overview, of these functional areas.

This analysis or overview can be performed by a company on an "in-house" basis or by outside consultants. An example of such an overview is outlined on the following pages.

ESTABLISHING A PLAN

The typical questions and facts requiring answers to establish a preseason organization and plan follow:

Merchandising Organization

a. Is the organization staffed properly?

b. Are present roles and duties properly established and clearly understood?

c. Do the top and middle management people have the capacity to produce the necessary results?

d. Is training necessary? Where and to what degree?

Sales

a. Establishment of sales targets, organization and performance, possible alternate or additional markets.

b. Review of marketing calendars.

c. Participation in style adoption process.

d. Involvement in planning.

Merchandising Planning, Timing and Functioning

a. Organization and present responsibilities.

b. Line development; Line "story", markups, stockkeeping units, inventory turns.

c. Adequacy of planning calendars.

d. Timing of events.

e. Assignment of responsibility for decision making.

f. Relationship of sales, delivery timing and control of mix going on shelf.

g. Control of piece goods and piece goods inventory.

h. Design and adoption of products: methods, timing.

i. Early and adequate involvement of manufacturing in the planning process.

j. Product engineering.

k. Costing: reconciliation of actual vs. estimated costs.

l. Duplicate sample making and timely delivery.

m. Specification flow.

n. Other merchandising tools.

o. Merchandising training: on the job and one-on-one.

Manufacturing - (Fabrics, Cut and Sew)

a. Organization and relationship with merchandising.

b. Production control: fabric, cut and sew, contractors, scheduling and loading.

c. Throughput timing.

d. In-process control - standard hours, minutes or labor dollars.

e. In-process lot control.

f. Evenness of production flow.

g. Variable losses throughout: labor job switching, make-up, other sources.

h. Production reporting procedures.

Customer Service Analysis

a. Average number of shipments per order.

b. Late shipments.

c. Returns analysis.

d. Cancellation due to failure to ship.

e. Order processing - methods.

f. Information flow.

Finished Goods Inventory Analysis

a. Average level of inventory.

b. Seasonal variations in level.

c. Viability.

d. Turns.

e. Losses due to markdowns.

Fabric Inventory Analysis

a. Average level of inventory.

b. Viability.

c. Fabric delivery.

d. Turns.

e. Loss due to markdowns.

Management Information

a. Review all types of reporting procedures.

b. Are the computer or manual controls adequate to support an improved situation?

Preseason Planning Impact

The merchandising/line planning function affects almost every area of an apparel company's operations. Several days of this course easily could be devoted to its many facets.

The intent of this segment is to provide you with a greater degree of understanding of line planning's key elements and to emphasize the critical need for improved communication, cooperation, and working relationships between the merchandising and manufacturing functions.

Merchandising/product line planning is a key activity that converts product/market strategy into desired (target) volume and exerts enormous impact on successful merchandise control results.

It is the task of executing the product/market strategy and is at the heart of a successful fashion-oriented apparel firm's day-to-day operations -- from initial analysis of merchandising trends to placing a line on sale.

Merchandising and product line planning and control are more critical in apparel than in most industries because of seasonality, fashion timeliness, and the potential volume of new product/style introduction. Determining a product line's effectiveness frequently represents 50 percent of an apparel firm's merchandise control task.

Merchandising/product line planning must include techniques to support and channel the creativity of a company's design group, directing its efforts to the target already identified by the product/market strategy.

Among the line planning sub-elements are:

a. Product Line Management (PLM) - to utilize available data and targeted research to translate product/market strategy into tangible/salable products.

b. Merchandising Calendars - to define what must be done to develop each line, when and by whom.

c. Forecasting - to utilize available data from PLM and other sources to project initially how much will sell.

The realities of merchandising/line planning in a fashion environment must include uncertainty and risk.

One corollary is that uncertainty means risk.

Merchandisers in the fashion business seek to control and minimize risk in a number of ways:

a. By increasing the variety offered in terms of garment styles and fabrics, they minimize the risk of missing fashion trends or "hot items." They hedge their bets by increasing the number of bets; some are bound to win. Extending a line to cover all possibilities, of course, introduces other risks in terms of excessive inventories, end-of-season markdowns, increased manufacturing costs, diffused marketing and sales effectiveness, and heavier investments of time and capital. There is a point of diminishing returns.

b. By delaying the decisions on fabrics and styling as long as possible, merchandisers are able to act on the most current market information. The later a company commits itself to a style, the less risk it takes with that style in the marketplace. However, late decisions create other risks: samples may be late; manufacturing costs will probably increase; delivery dates may be missed. Reducing risk can be risky.

c. Many firms attempt to survive in a fashion environment simply by making rapid changes in response to new market developments. They do not try to outguess the market, but instead try to follow it as quickly as possible. Depending upon the accuracy and timeliness of their information, and how quickly they can respond, following a market can be a successful tactic for some firms in the fashion business. But in return they must accept shorter selling seasons, and limited opportunities to engineer savings in manufacturing. Fashion is a world of compromises.

THE IMPLICATIONS OF UNCERTAINTY

So, uncertainty is a fact of life for fashion manufacturers. It presents them with the risk of not having the right product to sell at the right time, or that of producing too much of the wrong products, or of buying too much of the wrong fabrics. Efforts to minimize these risks generally involve delayed commitment to produce something, more styles than necessary, or costly changes to keep up with a rapidly changing market.

All of these attempts to minimize total risk entail additional risks of their own and must be controlled. Failure by merchandising to recognize the full consequences of these responses usually produces problems for the manufacturing end of the business.

Their efforts to minimize fashion risk are necessary, of course, but they are also responsible for part of the problems that manufacturing faces in a fashion environment.

The rest of the problems stem from the failure of manufacturing executives to understand fully and respond to the conditions that merchandisers must observe.

So, achieving two-way understanding and communication is the only hope for profitable growth in fashion manufacturing. The situation is far from hopeless. Merchandisers can employ their traditional efforts to minimize risk and at the same time minimize problems for their manufacturing sources if they:

a. Establish detailed time-sensitive planning and follow-up procedures.

b. Apply disciplined follow-up to their planning activities.

c. Provide maximum flexibility to manufacturing, within the constraints of merchandising requirements.

d. Develop and maintain enlightened cooperation with all operating functions of the company.

e. Provide constant, timely, and accurate communications with other activities.

These steps not only will help the merchandiser ease the problems that fashion creates for his or her counterpart in manufacturing, but also they will help that merchandiser achieve the full potential of his or her own efforts to put together a salable line that will produce a profit for the company.

CHANGE AND FLEXIBILITY

After uncertainty, the second reality of fashion manufacturing is change. Change is the essence of fashion, in more ways than one, and it has vast implications for both merchandising and manufacturing.

In merchandising, planning and controlling the development of a line up to the production process require control systems that will accommodate changes in forecasts, material procurement, production lots and sizes, delivery schedules, etc. New items are added to the line and others are dropped -- frequently at the beginning of the season, and reorders keep things fluid as the season progresses. Availability of specific fabrics, colors, and trim items causes endless substitutions. The merchandiser's lot is not an easy one.

On the other side of the fence, the manufacturer is challenged even further by the many changes that fashion manufacturing entails. The production executive or engineer seeks to maximize plant efficiency through engineered production systems and workplaces utilizing specialized equipment. Change dramatically affects this process. The more things change, the more they cost to make.

Yet change is also a fact of life in the fashion business, and manufacturers must learn to cope with it. The way to deal with change is to develop flexibility.

STYLE VARIATIONS COST MORE TO MAKE

Variety causes change in the manufacturer's world. Every time his/her production line must deal with a different operation or construction detail, different fabric or trim item, or a different color, he/she is faced with a new set of factors affecting costs and productivity. A change in the shape of a pattern is relatively simple; a change in the length of a collar point or the placement of a pocket can mean several hours' downtime for a specialized machine that has to be reset.

Because of the uncertainty and risk attached to each style in the line, initial runs are short, and delivery schedules are tight. The faster response times required in fashion manufacturing are essential for marketing success, but they take their toll in terms of higher manufacturing costs.

Table 6.1 attempts to quantify the impact of fashion on manufacturing by showing three degrees of style variation and the operating conditions they create. All of these are related in some way to the key elements of uncertainty and risk, change and flexibility.

Table 6.1 does not attempt to provide figures for actual companies but merely suggests the contrast between three degrees of style manufacturing. A full appreciation for these contrasts will give the reader a better idea of the scope of the problem.

FUNCTIONAL EMPHASES SHIFT

As a company deals with more styles, its different functional activities shift in importance. Some of these changes in emphasis are obvious; designing and styling play a larger role as the degree of style variation increases.

To determine what the customer wants requires a competent design and merchandise staff which is in close touch with the market and has the knowledge and budget to seek out those products, fabrics, colors, prints, and styling detail that will sell in the market.

The ability to forecast and reforecast accurately, correlating sales estimates for different styles with sales bookings as they arrive, and to coordinate that information with procurement and production scheduling so that salable products are available when needed for shipment, requires a responsive planning and control system run by skilled professionals. The

TABLE 6.1

THE EXPLOSIVE IMPACT OF FASHION ON APPAREL MANUFACTURING

FACTORS	DEGREE OF STYLE VARIATION		
	LOW	MEDIUM	HIGH
Seasons per year	2	4	6
Lines per season	2	3	5
Fabrics			
Fabrics per line	2	8	8
Colors per fabric	4	6	8
Fabrics/colors per year	32	576	1920
Stockkeeping Units			
Styles per line	8	30	30
Pattern sets per year	32	360	900
Sizes per style	6	6	8
Colors per style	4	6	8
SKU's per year (Style X Color X Size)	768	12960	43200
Average weeks per season	26	13	8.7
Average units per cutting order	4000	1000	400

same requirements must be set to control inventories of fashion fabrics--not only to have them available when needed for cutting, but also to minimize quantities of uncut fabrics at end of season. The availability of accurate information and the ability to apply it on time become critical. The EDP Manager gains importance.

STYLE DEMANDS MORE

A gain in importance for one function, to meet the demands of more style, does not necessarily mean less prominence for other functions. As we shall see, more style demands more from everyone in an apparel company. It is important for everyone, particularly top management, to recognize this.

In some functions, even through the total commitment of resources does not change with the addition of styles, there are changes required in talent and tactics for continued success. Manufacturing must become more flexible to cope with style change. Some trade-offs must be made between highly engineered and specialized operations that yield lowest costs and the need to push a new style through a shop in a hurry. The manufacturing team not only must gear itself to the different environment, but also it must understand the response for the changes, accept the challenges with enthusiasm, and respond as efficiently as possible. In the process, it may have to retreat from some operating principles and philosophies that never have been questioned.

CONFLICTS ARISE

As more resources are committed to the needs of style manufacturing, some conflicts develop. The goals of various functions in the company appear to be opposed to each other.

For example, late commitment to fabrics and styles helps merchandising be more correct in the market. However, such delays may result in tremendous complications and higher costs to manufacturing if samples are to be delivered on time to show retailers when they want to buy the line. If samples are not on time, the styles will not be sold regardless of how market-right they may be.

The designer may design products with construction and detailing variations which add tremendously to the cost of manufacture. Yet the designer feels those details are critical to the appeal of the garment.

Commitment of fabrics to production orders should be delayed as long as possible from a planning standpoint, yet this delay results in greater style mix, shorter runs, smaller cuts, and higher costs in production. There is also a greater need for shorter turn time within the plant.

Success in the fashion industry requires developing consistently successful product lines while minimizing markdowns and inventory levels. While this is certainly true of basics as well, it has to be achieved in fashion while hitting a constantly moving target of style, color, and price in situations where timing is of the essence.

Therefore, fashion-oriented companies must develop

a. A clear marketing strategy.

b. Effective merchandise planning and controls.

c. Flexible, cost-effective manufacturing resources with short lead times and rapid responses.

A successful fashion company must set up an environment, organization, and procedures where close cooperation and communication will lead to the understanding by each function of the critical needs of the other functions.

Then the effects of conflicting objectives can be minimized in a spirit of give and take, and overall objectives can be accomplished best.

ORGANIZATION AND OBJECTIVES

The whole process from line inception and planning through manufacturing and delivery is a continuum, as illustrated in Figure 6.1.

While manufacturing and distribution are self-descriptive and familiar terms, the interrelationships among marketing, merchandising, and sales are less clear cut and need amplification.

Marketing

Marketing is the advance process of defining all aspects of the market to be targeted, the overall strategy, commit-

MARKET STRATEGY

- Target Consumer
- Identify Market
- Assess Competition
- Determine Positioning
- Establish Pricing

MERCHANDISE PLANNING AND CONTROL

- Product Development
- Volume
- Timing
- Inventory Management

MANUFACTURING

- Competitive Resources
- Flexibility
- Timing

DISTRIBUTION

- Order Allocation
- Customer Servicing
- Inventory Disposal

FIGURE 6.1. PLANNING FLOW

ments, and advertising necessary to achieve merchandising and sales objectives.

Merchandising

Merchandising is the process which translates these objectives into specific product lines. It is responsible for their design, development, and promotion. Initiating the necessary plans and controls to make the line available at specific delivery periods is an important element of this function.

Sales

Sales is the implemention process of physically selling the line to the retail customer in acordance with objectives determined by the marketing and merchandising functions.

Although these functions are sequential, the degree of interface and reporting relationships vary according to the organization structure of the company involved.

7

Merchandising Planning, Forecasting and Controls

INTRODUCTION

Forecasting in a fashion environment to some degree reflects the activity of "high roller" gamblers. Forecasters rely upon instinct, subtle indications of change, copying, varying, and blind luck. Most firms recognize that forecasting is gambling and therefore build merchandise planning procedures to control their possible losses or to exploit rapidly favorable sales trends.

In translating marketing objectives into specific product lines, the merchandising function organizes its activities into a series of sequential programs.

a. Product identification to identify the product styles and categories that will result in maximum sales.

b. Product development creates a range of prototype styles from which to select the final line.

c. Product sales preparation to finalize all details on which sales budgets and production programs will be based, and make all preparations leading to the sales meeting and sales effort.

TIME-PHASED MERCHANDISING PLAN

For all the pieces to fit together exactly at the right time, a time-phased merchandising plan is needed. Such a plan lists all the activities necessary to identify and develop a product line and prepare it for selling, according to the sequence in which they must be performed and within the time frame of a season. The overall time available is established in accordance with marketing strategy decisions. Timing necessary for each activity and their necessary sequence (critical path) are provided by merchandising and manufacturing.

Establishing a merchandising calendar basically involves the following:

a. Establish the overall activity time frame considering external and internal constraints.

b. Develop a detailed schedule of all required activities and prepare descriptions of activities.

c. Determine activity sequence.

d. Determine activity start and finish dates.

e. Perform network analyses to determine the critical path of activities which will control the overall time frame.

f. Test feasibility of shortening the overall time frame by reducing time requirements along the critical path. Relaxation of time requirements for activities not on the critical path should also be considered.

g. Determine responsibility for completion of each activity.

h. Examine overlap of activities in each respective season and the implications on workloads at various times in the year.

i. Develop documentation, forms, progress reporting procedures, and controls.

Let's examine examples of a time-phased merchandising plan, and review each of its components.

PRODUCT IDENTIFICATION

Merchandising accomplishes this by means of market research. The research involves attending fashion shows, reviewing what fabric suppliers have to offer, looking at recent technological innovations, reviewing color and fabric libraries provided by fabric data services, visiting stores and outlets, and evaluating the opinions of buyers. The opinions of the trade press and market data services are constantly reviewed. Competitive products are purchased and examined. To identify the apparel product styles and concepts that will sell, the merchandiser must capture the pulse and direction of the market.

Part of this process involves evaluating how current styles and fabrications fit the emerging picture, and how many may be carried over into a new line. In some firms styles which fall below 1.5% of sales are dropped from the product line. Experience proves that fashion trends have a predictable degree of consumer acceptance over time. Once a product has been accepted, a period of growth will follow. At some point, however, consumer acceptance will peak, the style will go into a gradual decline, and finally will reach obsolescence. All products have their own particular life cycles. Many firms limit the total number of styles and sizes that may be offered. This forces weak styles and sizes out.

The actual positions of existing products on their life cycle curve is determined by analysis of current company experience at retail in terms of sell-through, bookings, and unsold merchandise, with similar analysis of retail store experience affecting the competition.

While this is taking place, the designer becomes involved and initial material purchasing decisions will be made. Two elements come into play here -- garment styling and fabric styling. Which influences which is a "chicken and egg" situation.

In some cases, fabric styling stimulates garment design. In others, certain garment designs require specific types of fabric.

Regardless of whether garment styles or fabric styles lead, practical considerations are dictated by fabric availability which must exactly fit the line development and production timetable.

Fabric availability lead times depend upon whether stock colors and designs are used, or whether specially designed fabrics are needed. They are also affected by the type of dyeing process used: stock dyed, yarn dyed, or piece dyed fabrics each have their own time cycles. Another issue concerns the amount of fabric testing required in order to meet the company quality standards.

Because of these variable factors, fabric orders must be committed at an early stage in development, possibly before design prototypes are finalized. Normally textile firms have long lead times which place the burden of risk on the apparel firm. For these reasons fabric purchasing is usually the responsibility of merchandisers who are held accountable for excess piece goods or finished inventory.

Product identification is completed with a detailed plan that outlines the styling concepts of the line, the fabrics and colors to be used in each style classification, how styles and fabrics are to be coordinated with each other, the size ranges required for each group of styles, and the types of accessories needed to supplement the lines.

The various elements of the product identification program constitute the first 16 activities on the typical time-phased merchandise plan shown in Figure 7.1 (From an AAMA example.)

PRODUCT DEVELOPMENT

In responding to the requirements of the merchandiser, the designer must be capable of producing a wide selection of prototype styles that are both innovative and adaptable to full scale production methods. This means that he or she must understand the cutting and sewing processes used by the company, their limitations and possibilities, and fabric properties and utilization.

There are limits to this, however. The designer's primary job is to create designs. Manufacturing management must be innovative and capable of adapting new designs and ideas into existing processes efficiently. Manufacturing performance is its prime responsibility. Each has constraints to observe.

To meet both requirements in a multistyle situation, it is essential to balance design requirements with manufacturing practicalities at the pre-production stage.

This process takes place during the interactive phase of product engineering, carried out by engineering personnel that are either part of, or who work closely with, manufacturing engineering.

During the product development stage, product engineering is responsible for evaluating the construction features of

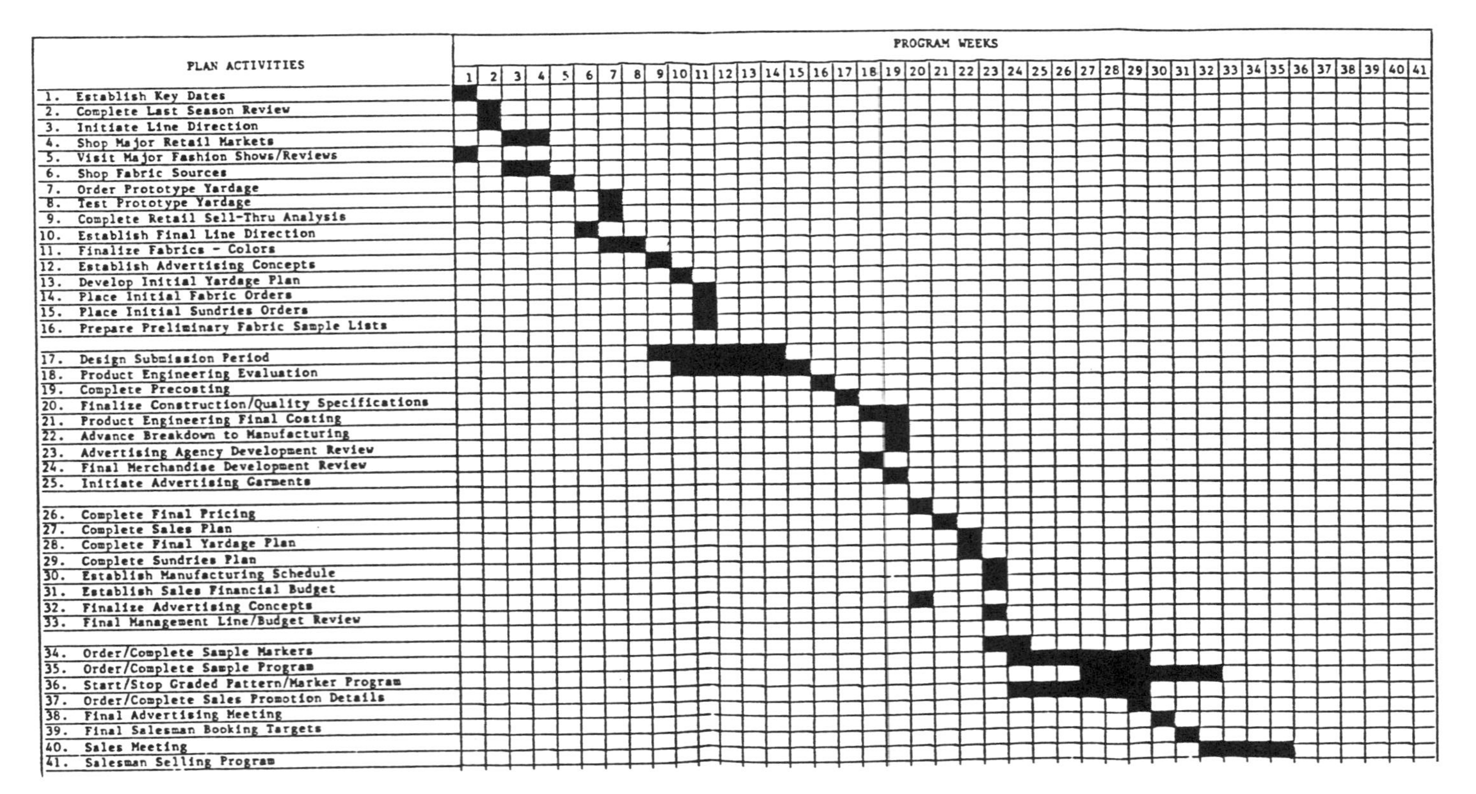

FIGURE 7.1. TIME-PHASED MERCHANDISING PLAN

new garments, reviewing the required sequence of operations, and evaluating those operations that do not fit current production methods and techniques. More often than ever today these engineers utilize computers to cost garments. Using a constantly updated engineering standards data base (either manual or computerized) provided by manufacturing, they estimate comparative time standards and cost alternatives for standard operations and those nonstandard operations required by new features.

In arriving at final solutions to design/manufacturing conflicts, some elements of negotiation and compromise are necessary. Provision of cost comparisons between various alternatives is an important part of the process. The most successful fashion apparel companies have achieved a high degree of effectiveness in this vital interface activity between the merchandising and manufacturing functions, creating an environment of mutual understanding, knowledge, and teamwork that maximizes overall operating performance.

During the follow-on reactive phase of product engineering, when construction and quality specifications have been finally determined, merchandising is provided with final labor and material costs for pricing and with material usage standards to develop the yardage plan.

Manufacturing is provided with advance operation breakdowns and labor cost standards. Used in conjunction with the merchandise style/unit estimates, manufacturing can prepare pre-production plans for advance plant loading, equipment ordering, and assessment of retraining needs.

As design prototype garments are developed, they are submitted to a continuous planned review process involving both designer and merchandiser. Cost information from product engineering assists this evaluation.

The extent of such review activities is usually governed by the number of lines in preparation at one time, and the time cycle allowed by each. In companies that produce multiple short cycle lines each year, design creativity can be channeled more effectively by narrowing down the number of prototypes. This factor affects the timing and scope of the product engineering activity.

The need to adapt new designs efficiently to existing conditions remains constant, with product engineering providing a valuable service function to aid merchandising.

The various elements of a product development program include activities 17 through 25 on the time-phased merchandise plan shown in Figure 7.1.

PRODUCT LINE SALES PREPARATION

When line content is finalized, merchandising then projects the units estimated by style and fabric to develop:

a. A sales plan showing total units by style and fabric.

b. A final yardage plan which shows total quantities required by fabric type.

c. A similar plan covering trim and sundries.

d. A sales (financial) budget projecting the profit margins for each style number, each major product group, and overall seasonal line.

e. Advertising and promotional concepts, and the selection of garments for promotion by the advertising agencies. (These programs are successively reviewed during follow-on agency presentation meetings).

f. A final line review by general management and merchandising in which all plans are approved and committed.

g. Salesmen's sample style defined and ordered, based on the sales plan.

h. Sample patterns and sample markers.

i. Graded patterns and, if applicable, production master markers.

j. Sales promotion brochures. Sales promotion areas are targeted and salemen's booking goals are established.

k. A sales meeting to show the line, explain sales policies, provide direction, outline pricing strategies, and review sales support programs. (At this time, salesmen would formally forecast their expected sales by style and fabric.)

Depending upon the timing of production relative to sales, a sales plan could involve a smoothing of the style mix and unit quantities assigned to each salesman and could lead to a redistribution within their total sales within their total sales allocation. Sometimes the plan would permit revision in styles and fabrications not yet produced. This action would constitute the first adjustment to the previously established plans and schedules.

The various elements of the product line sales preparation program include activities 26 through 41 on the time-phased merchandising plan in Figure 7.1.

That example has been limited thus far to showing all the different planning activities necessary to develop one line for one season. A company with a high degree of style variation may have six selling seasons per year, with perhaps five lines for each season. In such a situation, there will be seasonal overlap at times during the year. Merchandise planning activities for different lines and seasons might be carried out at the same time, and the need for some way to organize all of them becomes apparent. Figure 7.2 (from an AAMA Systems Report) shows the impact of a higher degree of fashion on a typical time-phased merchandising plan. Each block series indicates a different season, and the scope

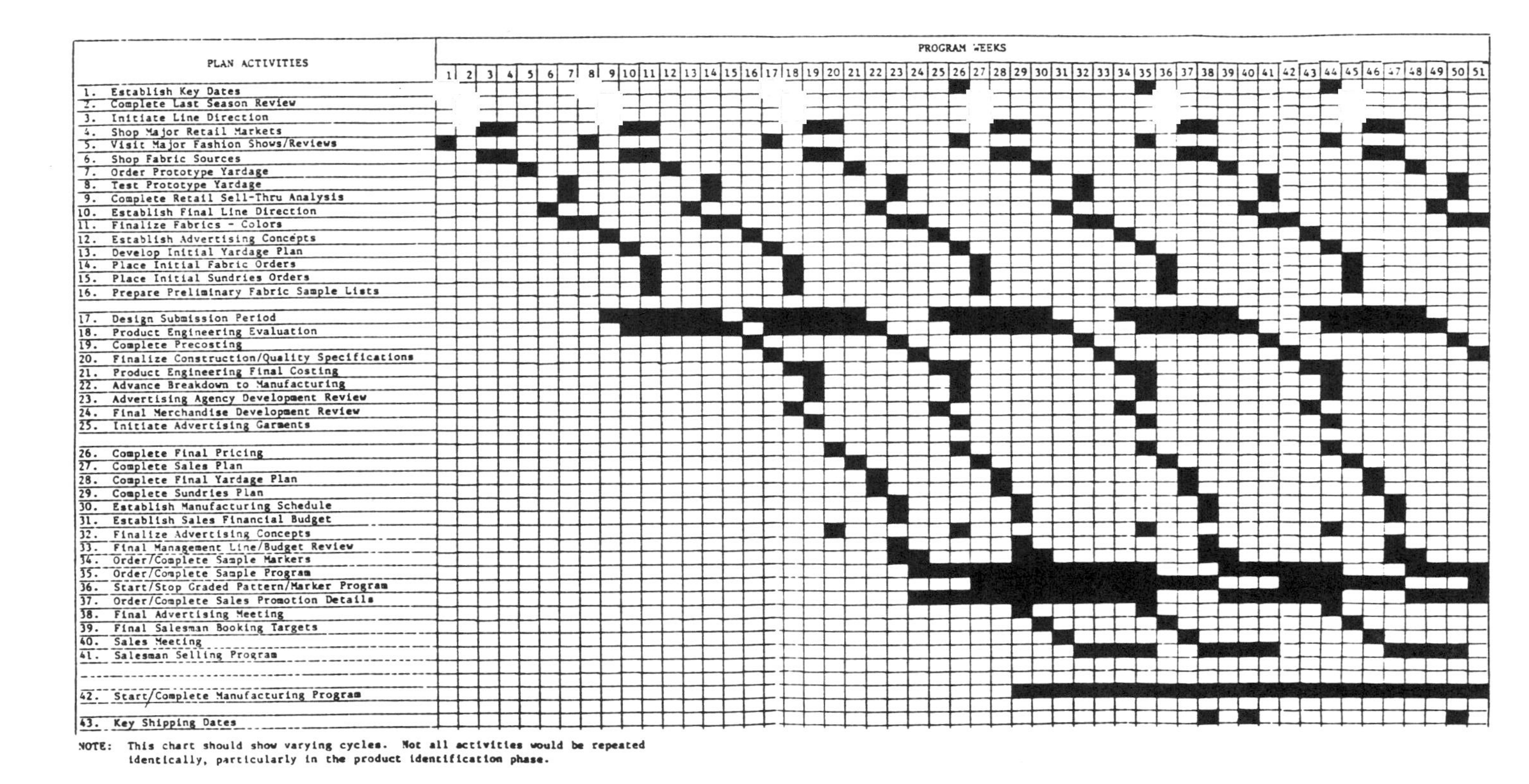

FIGURE 7.2. TIME-PHASED MERCHANDISING PLAN—MULTISEASON

of merchandise planning activities explodes as lines are added in each.

As noted in Table 6.1 in the preceding chapter, merchandise planning in a high fashion environment can encompass close to 2,000 fabric/color combinations per year, as many as 900 sets of patterns, and over 43,000 style/color/size combinations.

THE NEED FOR DISCIPLINE

It must be emphasized that discipline is essential to organize and control the entire merchandise planning process. As style variations progress from low to high degree, the demands made upon a fashion company's planning capacity increase exponentially--as do the number of uncertainties and risks faced by merchandising.

Managing style at this level becomes increasingly complex, and more discipline is needed. Intuitive feel for style, fabric, and market trends -- the highly individualized creativity that is traditionally associated with fashion -- is gradually replaced by organization, system, and disciplined planning. Fashion is big business today.

Depending upon the size of the company and its degree of fashion involvement, such control can be carried out by merchandise planning with overall responsibilities to keep all activities on schedule. In less complicated situations, the same results can be achieved through less formal but regularly scheduled meetings where merchandising and manufacturing representatives can exchange information and keep their activities coordinated.

Regardless of the format, some form of time-phased planning for all merchandising activities must be established. If the responsibilities of merchandising are clearly stated and are accompanied by deadlines that will leave enough time for manufacturing to respond, the complexities of fashion can be controlled. Again, more discipline on the part of merchandising is the key--and it is up to top management to ensure that the necessary degree of discipline is present continually.

Merchandising executives and their staffs must be convinced that their job performances will be measured not only by the success their lines enjoy in the marketplace, but also in terms of how consistently they meet the deadlines of a time-phased merchandise plan.

Through the use of such plans, they should also come to understand that the execution of their styling and merchandising concepts will be significantly more effective if they remain on schedule. Given more timely and reliable information, manufacturing will be able to do a better job of carrying out merchandising's plans.

Lastly, it is essential that market needs come first, and whether the manufacturing structure in place is all internal, external, or offers some of each, existing capabilities and/or capacities of manufacturing should not be permitted to limit or constrain the potential of the product line that has been determined by top management, marketing, and merchandising to be correct for the company's market strategy.

PRESEASON FORECASTING AND FORECAST CONTROL

An integral component of a time-phased merchandising plan is a preseason forecast of projected season sales by style and, where applicable, by color. It is also a key component of an apparel company's inventory management system, in that it represents a point at which significant commitments are made of a company's resources in terms of raw materials, findings, and sundry supplies.

The development of an initial preseason sales forecast is a key merchandising responsibility after completion of line development and finalization of line structure. It must initially reflect two forecast projections for each color/style. The first projection is a total season estimate; the second is an initial buy plan (initial fabric ownership) against the season estimate.

While merchandising management must rely on its judgment to a great extent, initial forecasts, made in concert with marketing and sales management input and frequently reviewed by top management, need not be based entirely on intuition or "seat-of-the-pants" methods.

Even in a fashion-oriented environment, there are identifiable trends, cycles, and distribution patterns which can help to improve sales projections. Some progressive firms are now utilizing this type of data in mathematical models; others are helping themselves with basic, but carefully designed and well-organized, presentations of historical information.

Formal "initial buy plans" must be coordinated with preseason manufacturing plan (including assignment of styles by plant or external sources), and this integration should be used to control and schedule raw materials and production that will support the initial forecast and meet anticipated delivery objectives.

The combination of more frequent and shorter seasons, substantial preseason fabric commitments, significant variations in construction, and the needs to ship coordinated garments or groups (e.g. Misses' sportswear) virtually demands detailed preplanning of purchases and production on a time period basis.

In a proper system, the first rough buy plan and manufacturing plan are set up as soon as an initial overall forecast is established. These plans should be revised as forecasts are updated and carried to more detailed levels. Working with only seasonal grand totals by fabric/pattern and production requirements, for example, often results in unbalanced inventories of both materials and finished goods.

An example of a seasonal forecast control report format (non-time-phased) is outlined below. This format, or modifications thereof, can be used in a manual or computerized system. It is essentially a summary management/ merchandising control report that can be used for:

a. Initial season forecasts.
b. Forecast monitoring.
c. Reforecasting.
d. Inventory management analysis.

forecasts and reforecasts to maximize inventory turns and optimize the utilization of manufacturing facilities, whether internal or external, to the degree permitted by inventory management targets. The time-phasing of initial forecasts by required delivery periods provides for the effective comparison of merchandise needs vs. manufacturing capacities by delivery periods, and results in the most effective trade-offs between customer delivery requirements, inventory management targets, and manufacturing facilities utilization.

In an era of high interest rates, apparel industry firms are increasingly moving away from the industry's manufacturing orientation (capacity analysis) to a market and inventory management focus which can and has affected the degree to which manufacturing facilities are utilized on a year-round basis. Time-phasing is essential to accomplish the best balance between the potentially conflicting objectives of merchandising and manufacturing and to achieve the financial objectives of the company.

Figure 7.3 illustrates an example of a time-phased seasonal master forecast planning report format which can be used for:

a. Time-phasing original season forecasts.

b. Time-phasing season reforecasts.

c. Time-phasing balance of season ownership (in season).

EXHIBIT ABC INDUSTRIES, INC.
CASUAL SLACKS
SEASONAL FORECAST CONTROL REPORT

Last Report Dated 1/16/82
This Report Dated 1/30/82
Season Spring 1982

						Total Order Status			Total Production Status						
Lot No.	Customer	Color	Season Estimate	Units On Contract or Blanket	Season Ownership	Units Ordered	Units Shipped	Units Unshipped	Cut STD	Shelf Stock	Cut and In-Process	Issued Uncut	Unsold Cut Position	Unsold Ownership Position	Comments
333	J. C. Penney	Charcoal	125,000	80,000	80,000	40,000	15,000	25,000	50,000	12,000	17,000	6000	10,000	40,000	
337	J. C. Penney	Brown	10,000	6,000	6,000	3,050	1,115	1,935	4,100	935	2,050	--	1,050	2,950	
		Navy	14,000	9,000	9,000	3,900	1,470	2,430	5,200	1,130	2,600	--	1,300	5,100	
		Grey	16,000	10,000	10,000	5,050	1,615	3,435	6,700	1,735	3,350	--	1,650	4,950	
			40,000	25,000	25,000	12,000	4,200	7,800	16,000	3,800	8,000	--	4,000	13,000	

This report can be used for

(1) Original Season Forecasts
(2) Forecast Monitoring
(3) Reforecasting
(4) Inventory Management Analysis

FIGURE 7.3.

d. Time-phasing balance to season estimate (in season).

MASTER SCHEDULING AND CONTROLS

Master scheduling is the critical linkage among forecast, actual, demand, delivery objectives, raw material availability, and capacity availability. A master schedule is the only realistic way to deal with a complex information plan in today's apparel company, allowing it to react to shifts in demand and materials to achieve objectives in delivery and inventory turns.

An up-to-date master scheduling system includes modules for production planning, material planning, capacity planning, cut issuing and factory loading, and work-in-process control. These provide the hard copy link between merchandising and manufacturing. Current computer technology techniques can significantly upgrade what can be achieved in this area in terms of timeliness and accuracy.

A master schedule, or forward production plan, covering an entire season or forecast period will, in its final format, beat the style/color level of detail and will be broken down into time increments. In many firms, products are reviewed against all production lines to optimize the use of manufacturing facilities and to minimize changeover costs between seasons.

Because the master schedule is the foundation for all subsequent planning and scheduling activities, it must be continually updated to keep it in line with revised forecasts, delivery objectives, raw material availability, and manufacturing constraints.

Next to accurate sales forecasting and forecast correction, the maintenance of a detailed, time-phased master production schedule (seasonal manufacturing plan) is the most important element in an inventory management system. It is fundamental to the development of sound raw material requirements and capacity planning systems and to the establishment of reliable finished goods delivery schedules.

Apparel firms must ask themselves the following questions:

a. Is our production and raw material planning system based on a time-phased style/color manufacturing plan which is continually updated to keep it in line with current forecasts, delivery objectives, raw material availability, and manufacturing constraints?

b. Do our planning reports facilitate management review of the current plan in terms of its effects on raw material and finished goods inventory requirements and customer deliveries?

c. Does our plan provide the primary basis for fabric and trim re-buys, production level adjustments, piece goods delivery follow-up, and the issuance of cuts?

Many apparel firms, some with substantial data processing support systems, continue to plan and buy solely on the basis of cut and sold totals, ownership vs. forecast, etc. They reject the concept of detailed forward planning because of the difficulty of maintaining the plan in the face of frequent forecast changes, plant production problems, and piece goods delivery failures.

This is not a valid argument. For apparel firms large and diverse enough to need them, there are computer techniques available which will do most of the detail work. In addition, if a company is going to have a "production planning" function, those individuals should, in fact, spend their time planning, not "fighting fires." In many companies, the so-called production planning group spends much of its time and energy dealing with each day's crises. Without a reliable system, "Planners" cannot plan; they only react to unplanned obstacles.

An integrated information system helps minimize (not eliminate) these problems and enables management to deal more effectively with those that remain.

A schematic diagram which portrays the major activities/decisions affecting the levels and makeup of inventory before and during the season is displayed in Figure 7.4. To the extent that an apparel company's management information system integrates these activities and initiates informed, timely decision making, that company will be in a position to manage its inventories more effectively.

An example of a fabric forecast control report format, an element of an integrated management information system, is illustrated in Figure 7.4.

A master production plan control report format, illustrating an in-season reporting activity, integrating time-phased uncut "ownership" and planning (not-owned) season estimates with a "capacity analysis," is illustrated in Figure 7.4.

EXHIBIT ABC INDUSTRIES, INC.
CASUAL SLACKS
FABRIC FORECAST CONTROL REPORT

Projection Basis:
Master Forecast Planning Report Dated 1/30/82
Last Report Dated
Season Spring 1982

Lot			Season		Cut and	Purchase	Undelivered Fabric											
					Issued	Order	Fabric			On-Hand	Due Yards						Total On	
No.	Customer	Color	Estimate	Ownership	STD	Number(s)	Mill	Pattern	Color	Yards	January	February	March	April	May	Later	and Due	Action
333	J. C. Penney	Charcoal	U 125,000	U 80,000	U 50,000	137,891	Burl	P/7639	Slate	--	--	30,000	15,600	--	--	--	45,600	
			Y 165,000	Y 105,600	Y 60,000													

FIGURE 7.4.

8

Materials Scheduling and Utilization

INTRODUCTION

Preseason planning represents the "give and take" between sales and manufacturing. In the section on preseason planning, it was pointed out that plants may be preloaded for labor and equipment to optimize plant use and to minimize operating costs. Most production planners "bulk load" plants using known equipment or skill bottlenecks to control the load. Once this is done, detailed day by day loading usually follows at the plant level. Computers also calculate other elements in detail loading. Using a materials requirement planning (MRP) system, a production planning department explodes the product specifications for material and trim into quantities needed. In order to achieve reasonable accuracy in ordering raw materials, a system must be in place which is capable of providing raw material usage (waste, etc.) estimates. In some firms this function falls under a material standards control unit which follows usage of cloth, trim and thread weekly by plant. Preseason and inseason planning is part of an integrated management information system (see Fig. 8.1).

MATERIAL STANDARDS CONTROL

MSC clerks estimate the quantities of materials required to produce a garment. This estimate is based on a garment review. The clerks then enter their estimates by style in the computer. Normally MSC clerks review the merchandiser's list of items which comprise the product structure for the new style against a prototype garment. If this list is complete, the clerk, using past experience, estimates the item quantities required. Item volumes, such as piece goods, are dependent upon the sizes offered and the average size to be sold. (These records are usually continuously maintained in the computer from prior seasons.)

The estimator normally prepares a miniature marker (either on a computer or manually) to verify the piece goods yardage allowance. Estimates may be revised as experience with a new material indicates the need. Where a new fabric requires more than estimated usage, losses can be substantial if correction in material utilization and product cost is not promptly made during the production of the garment. Errors can also lead to product shortage and lost sales if left unadjusted.

MATERIAL REQUIREMENTS PLANNING

Material requirements planning (MRP) is a 20 year old computer program which is a relatively new solution to an old problem for apparel manufacturers. The system is

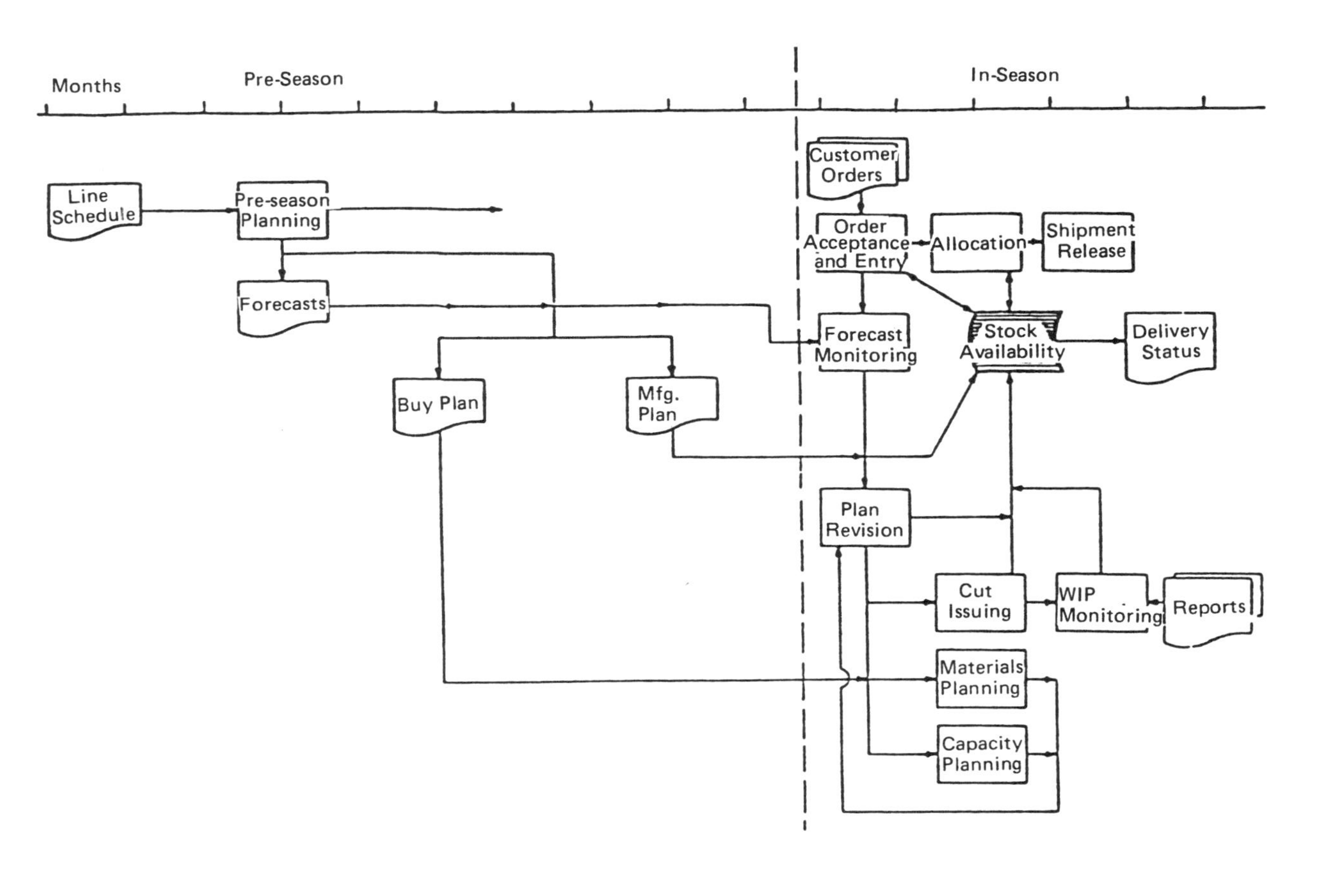

FIGURE 8.1. AN INTEGRATED MANAGEMENT INFORMATION SYSTEM

designed to have material on hand when needed without carrying excess inventory. MRP is the art and science of using a computer to collect and classify data in order to solve problems in production control and inventory management. With MRP, complex interrelationships can be resolved relatively rapidly. For example, suppose that item A (i.e., orange yellow thread) is used in several end products and that the demand for these end products is changing constantly. How does this affect demand for item A? Before installing MRP, the average manufacturer usually carried a heavy inventory of item A and hoped that it would be adequate. With the MRP system, each change in end product demand can be instantly reflected into inventory requirements. Production planning can then either reorder or hold back on item A orders well before the final production occurs.

MRP is not a fixed body of knowledge. The system's use continues to expand as organizations become more adept at using computers and reviewing data. The term was first widely used in the late 1960s. At that time material requirements planning pretty consisted of using a computer to perform operations that had previously been done by clerks. The system also permits use of time-tested simple statistical techniques such as: economic order quantities; application of probability theory to determine safety stocks; transportation matrices and statistical demand forecasting.

As a matter of fact, these techniques have been available for years. They were not in wide use by apparel firms simply because the calculations needed to make them work were too time consuming and tedious. As MRP use grew, the system has become more sophisticated and uses more complex models that approach closer to real life. The location of MRP in the planning flow is illustrated in Figure 8.2.

The basic goals of MRP systems include:

a. Least cost protection against uncertainty. Organizations tend to build stockpiles; manpower, machinery and materials as protection against an uncertain future. Management systems cannot eliminate the unexpected (such as a strike at a supplier plant or a shipping delay). Systems are built to rapidly transmit adverse information for immediate evaluation. Most conditions are normal. MRP makes it possible for organizations who maintain proper historical data to operate with less allowance for uncertainty. As a result, inventory, facility and operating investments and costs can be reduced.

b. Optimum balance between labor and capital investments. Almost all economic activity is carried on with some mix of labor and capital. To a limited extent, one production factor can become a substitute for the other. For example, capital and labor can be a substitute for each other. Two examples are:

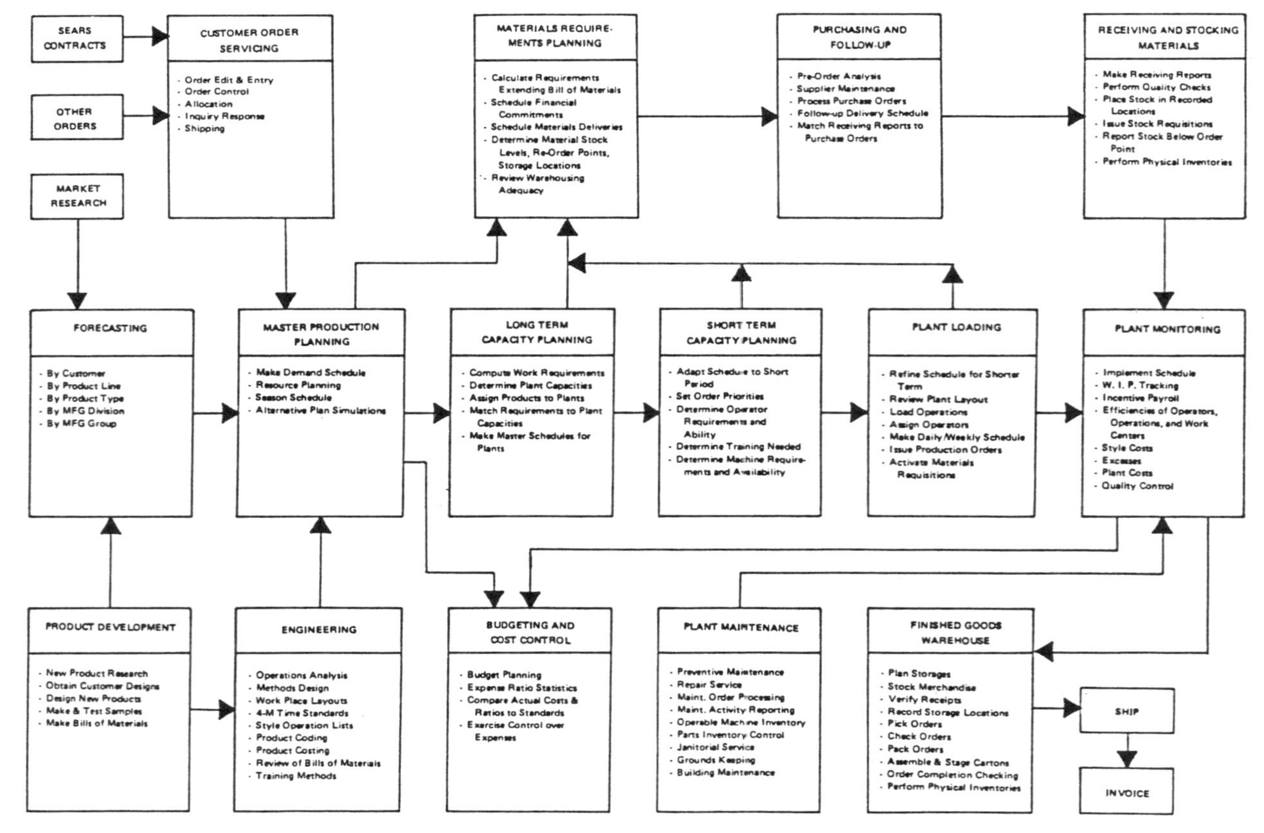

FIGURE 8.2. MANUFACTURING OPERATION SYSTEMS FUNCTIONAL FLOW BETWEEN MODULES

(1) A company forecasts a product demand at 1,000 units for a season. Should the firm manufacture in lots of 100 or in one lot of 1,000? With the larger lot, more capital is tied up in inventory. The risk of obsolescence is increased. On the other hand, the labor and equipment component costs will probably be less if the larger lot is produced.

(2) Raw materials stockpiles support production. The inventory size and cost are related, in part, to the frequency of stock level review. Frequent review provides more opportunity to reorder; therefore, investment can be reduced. Computers permit more frequent reviews. A good MRP system permits random physical inventories.

c. Use of labor and capital. Most apparel production processes consist of a series of events in which materials are waiting in line to be processed. The queues can be shortened by altering the labor and/or capital inputs. Longer queues may permit the organization to get along with less labor or capital but only at the price of in-process inventory levels. MRP can be used to minimize capital, labor, and other in-process costs.

MRP SYSTEM ORGANIZATION

Material requirements planning in a manufacturing organization is normally carried on by the production and inventory control department. Production control is the process of monitoring the performance of workers and equipment. Monitored data is compared with a predetermined plan. The production controller's function is to see that the objectives of the plan are being met. The department should also evaluate how efficiently human and physical resources are being used (see Fig. 8.3).

Nonmanufacturing Applications

Production control functions are not limited to manufacturing. Production control is universal to all economic activity, although it cannot always be easily identified. The traffic department of an airline or railroad is engaged in production control as the firm decides what equipment to have on hand in a particular location. The registrar of a university is engaged in production control when he or she determines that a particular classroom will be used for a particular course at a particular hour. The manager of a retail store is engaged in production control when he or she determines that eight salesclerks will be needed on Saturdays but only six on weekdays.

Production control is concerned with the utilization of resources (labor, equipment and materials) needed to get a job done. The production control manager is concerned with providing the operations manager with the necessities to get the job done.

FIGURE 8.3. PRODUCTION CONTROL AND PRODUCTION PROCESS

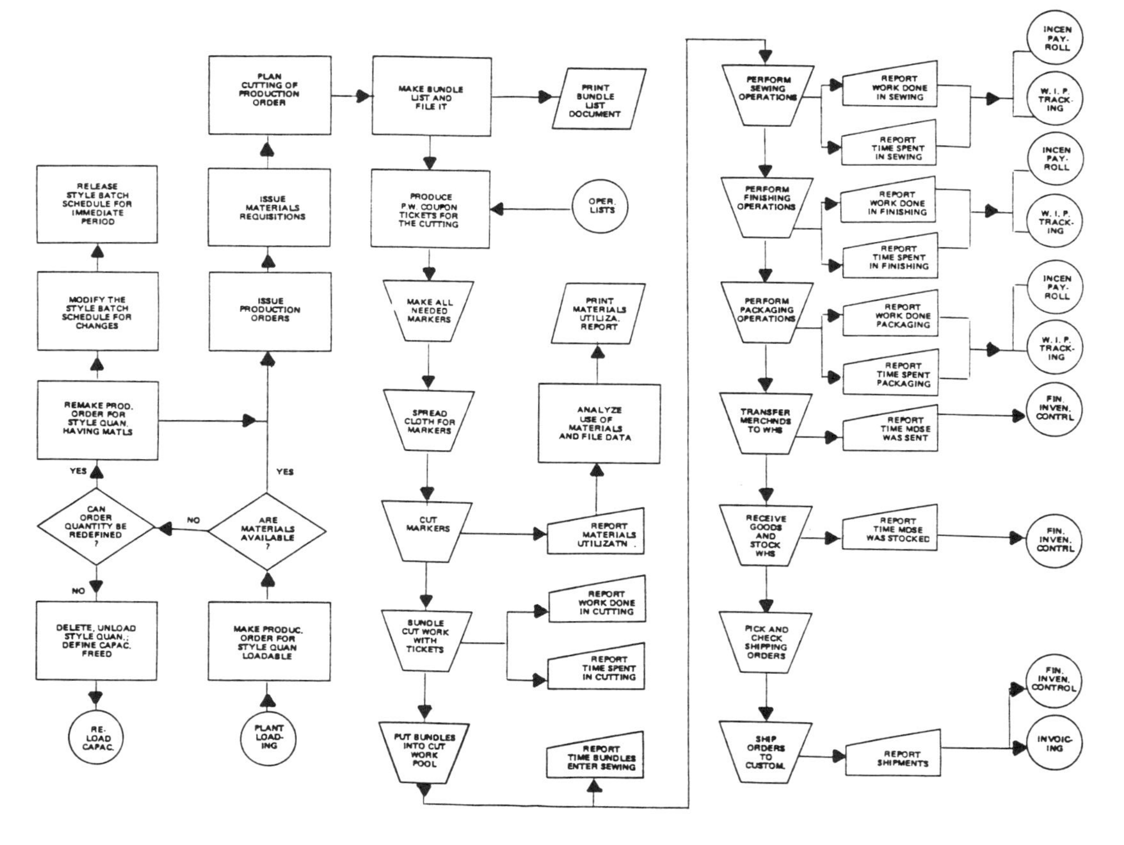

Materials Management Function

How, then, can one justify production control as a part of materials management rather than the operating management activity itself? There are basically four reasons.

a. Of the three resources used in production, material is the only one that is external to the organization. Over the short term, the labor and equipment are already there. Long term planning was covered in the chapter on corporate goals. Production control primarily plans materials. This activity consists of matching material, workers and machines so as to meet promises to customers.

b. The skills required for materials management are transferrable to production control problems involving labor and equipment.

c. Production control problems involving material, labor, and equipment are interrelated. The basic production control problem is to bring all three resources together at the right time in the proper quantities.

d. The materials management organization can be a part of operations management. When operations management embraces materials management, production control can be part of both, with no real conflict. When production control is part of a separate materials management department

> (that is independent of the manufacturing organization), its decisions involving the scheduling of labor and equipment must nevertheless be consistent with manufacturing's objectives, just as the physical distribution part of the materials management activity serves marketing's objectives. Otherwise there will be conflict.

The production control process is both easy and difficult to visualize. In small organizations it is carried on so inconspicuously that it does not seem to exist. In large organizations its existence is all too apparent, and the process becomes so complex that it is difficult to comprehend.

In the small organization, for example, production control and MRP consist primarily of a worker asking the boss which job to start on next and, from time to time, the boss checking up and asking the worker how he or she is coming along. This is more or less what happens in a very large organization, but here there may be thousands of workers and hundreds of bosses.

In large organizations oral instructions are no longer adequate; in fact, there are so many activities going on simultaneously in various plants that it is difficult for any single person to grasp what is happening. The total report of whether or not everything is going according to plan usually is determined by comparing a computer stored production plan to production activity data flowing from various production lines.

MATERIAL UTILIZATION - THE BASIS FOR PRODUCTION INVENTORY CONTROL

The Role of Fabric Costs

It is generally accepted as fact that fabric costs represent some 40 to 60 percent of the wholesale selling price of most mass-produced apparel products. A 5 percent fabric savings can represent the equivalent of a 2 to 2-1/2 percent contribution to apparel corporate profit before taxes.

Because of this cost ratio, many apparel manufacturers have undertaken periodic material utilization programs over the years. This trend has been stimulated by the introduction of automatic miniature marking equipment and the use of video tubes to make markers. However, as an industry, apparel firms periodically lapse from proper control over raw material usage. Usually this occurs in trying to meet out of line promises to customers.

Material Utilization Controls

Typical manufacturers usually know the techniques for controlling material usage. Rarely do manufacturers use what they know in a total assault on wasteful material usage practices.

a. A recurring goal is develop one system for consistent evaluation of MU performance. This requires that performance be measured and evaluated regularly.

b. This goal is seldom attained. Most apparel companies are not organized to do this. Where MU controls exist, they are often superimposed upon normal manufacturing operations. Even when the control structure represents a well-coordinated application of known techniques, it often remains outside the daily interdependent workings of the organization. The control system is almost an afterthought.

A second part of this problem is that most MU controls generally represent an extra effort for the personnel in the affected areas. Thus the activity can result in extra costs or possible delays to their normal activities. MU controls usually run counter to the normal manufacturing pressures for increased production and lower labor cost. Because of this, control tends to deteriorate and eventually disappear. The advent of the minicomputer together with packaged MU programs improves a firm's opportunity to exercise tight MU control.

Computers can calculate the base usage required to produce a garment as in Table 8.1 below.

TABLE 8.1
PATTERN AREA SQUARE INCHES

	SIZE			
Part	32	34	36	38
Front	200	225	248	275
Back	239	267	301	334
Top Back	53	59	64	70
Sleeve (2)	138	159	180	192
Total				3,004

The computer then compares the need against marker use as shown in Table 8.2.

TABLE 8.2
TOTAL SQUARE INCHES OF PATTERNS IN THE TEST MARKER

Part	SIZE 32	34	36	38
Front	400	1350	1240	275
Back	478	1602	1505	334
Top Back	166	354	320	70
Sleeve (2)	552	1908	1800	384
Total				12,738

From this comparison will come the ability to evaluate a marker as shown in Table 8.3.

TABLE 8.3
TEST MARKER EVALUATION

Test Marker 45"x 8 yd. 17" (305") = 13,725

$$\frac{13,725}{12,738} = 107.7\% \text{ waste in marker}$$

$$\frac{12,738}{13,725} = 92.81\% \text{ efficiency in marker}$$

This calculation can be performed on most computerized marker systems. Generally, the marking supervisor sets loss limits for marker efficiency. The marker maker continues marking until the standard is met.

End Loss

An end loss calculation based on an expected one-inch loss at each end of the spread and an average marker length of twelve yards follows. The two-inch loss divided by twelve yards times 36 inches produces a .46% loss.

End Loss - 1" for each End

Average 12 Yard Markers

$$\frac{2''}{432''} \times 100 = .46\%$$

Some firms predetermine acceptable end loss figures. There are computer package programs available that will calculate end loss versus width loss versus spreading and cutting labor. Other forms of piece goods losses requiring monitoring follow.

Edge Loss

An edge loss example assumes that the markers will be 44 inches wide for 45-inch fabric. This produces a half-inch loss on each side. The total of one-inch loss divided by 45-inch fabric width produces the 2.22% edge loss.

Edge Loss - 1/2" Each Side

45" Fabric

$$\frac{1''}{45''} \times 100 = 2.22\%$$

Piece End Loss

An example of a piece end loss calculation; assume an average piece length of 37 yards; a pattern length of 20 inches, and an end loss of one-half inch. In addition, there is the two-inch lap back allowance. All together this adds up to a total loss of 12 inches at the end of each piece. The twelve-inch loss in every 37 yards is converted to a percentage by dividing the quantity 37 yards times 36 inches into the 12-inch loss.

Piece Ends - Piece Length - 37 Yards

Average Pattern Length - 20"

$$\frac{12''}{37 \times 36} \times 100 = 0.90\%$$

Defect Loss

Since cutting out a fabric defect produces a piece end, an end loss will occur each time a defect is cut out. As an example, if we assume a defect frequency of 5 per hundred yards, then the 12 inches end loss will apply. The calculation of percent waste can be made as illustrated.

Defect Cut Outs - Defects Allowed per 100 Yards = 5

$$\frac{5 \times 12}{36 \times 100} = 1.67\%$$

Total Fabric Requirements

Totaling the loss percentages in the above example will produce a number for calculating the amount of cloth to be ordered. The calculation of fabric requirements for each style can be made by increasing the marker standard by the sum of these additional losses. (Note: Other styles may have different loss percentages; i.e., multiplying the marker loss percentage to increase the actual pattern square inches.) This latter method is illustrated as follows:

Losses:

End	.46%
Edge	2.22%
Piece End	.90%
Defect	1.67%
Total Losses	5.25%

Marker Loss % = 7.75%

$$\frac{107.75 \times 105.25}{100} - 100 = 13.34\%$$

Additional MU Factors

Most apparel manufacturers are far more sophisticated and accurate in analyzing and measuring labor costs and production volumes than they are in evaluating material utilization performance, especially with style-oriented patterns and fabrics. Estimating appropriate yardages for purchase is further compounded by the firm's styling, merchandising, and planning decisions.

For example, the specification by designers of a fabric whose width is narrower than width normally used by the markers, or whose delivered width varies a fraction of an inch under that specified, can limit the ability of the cutting room to make full use of fabric. (This shortage of square inches available also may lead to the production of fewer garments at higher costs.) Equally limiting, but less often appreciated, are those merchandising and production-planning decisions which take away the flexibility of the cutting room in terms of using optimum cutting heights, marker lengths, size assortments, or pattern fits.

The Purpose of MU

a. The basic concept of an MU program is to:

   (1) Determine areas where savings can be achieved.

   (2) Design a program to make those basic savings.

   (3) Establish a complete series of controls to ensure that the savings are attained and maintained.

b. There are a number of areas in which there are normally potential savings. They are not always found or applicable in every company. Table 8.4 provides an example of savings converted from dollar sales volume.

c. MU improvements can be converted to piece goods inventory terms. Reducing piece goods inventory by installing a sound MU program can lower a firm's working capital requirement.

Consultants use the above criteria as a way to evaluate cutting department weaknesses. Deviations from the norm serve as system improvement attack points.
A properly designed MU program will use a computer to track material usage against usage standards. Some computer programs go as far as to specify which pieces are to be cut together in order to minimize width losses.

TABLE 8.4

NORMAL MATERIAL UTILIZATION IMPROVEMENT RANGES

Function	Normal % Improvement Range Minimum	Mode	Maximum	Normal Potential Savings Per $ Million Sales[a]
Fabric Inspection	0.0%	0.5%	2.0%	$ 2,000
Width/shade utilization	.0	.5	1.5	2,000
Pattern/marking rule engineering	.0	.3	3.0	1,200
Cut planning	.0	1.0	3.0	4,000
Marking controls	.0	1.0	2.0	4,000
Spreading losses	.0	.5	3.0	2,000
Recut controls	.0	.2	.5	800
TOTALS	0.0%	4.0%	15.0%	$16,000

[a]Assumes material cost to be 40% of sales price.

9

Production Control—Initiating Cutting

INTRODUCTION

Reference has been made to Production Control's role in seasonal planning. This department is also responsible for the process of carrying out plans and schedules already made. Execution of this activity begins with the authorization of production, continues with the regulation of the flow of work through process and ends when the finished merchandise is shipped. In many apparel firms responsibility for the accomplishing of these actions may be split between home, office, divisions, and plants. Regardless of how the results are achieved, computer systems are playing an increasing role in facilitating control activity.

As fashion and product variety increases, production control becomes more difficult to execute in the plant. As time passes and apparel buying activity changes, every kind of apparel production process is affected by product mixes containing dissimilar products. This trend creates short production runs. In-plant work loads shift continually. Shortage of some operational skills develop as does the lack of the suitable equipment at some points in process. Because of the continually changing mixture of products in process, the bottlenecks may shift to different places within the plant daily.

Production control becomes more complicated as the need for flexibility increases. The first step toward coping with change is to establish a system that accepts change as a normal environment. Stability in a production process must be recognized as a theoretical concept rarely realized in a plant.

Supervisors must be trained to accept changing mixtures of products. Operators must be able to accept changes of operations and machines. Mechanics must be ready to change equipment or attachments at work places. Fashion requires the ability to follow and to cope with change. Computers can be used to predict production problem points.

Systems of production control must provide for determining from data what the status of the production process really is. It should be possible to compare the state of plant activity to a plan. Places in which trouble may occur can be made detectable and so that there will be a basis for deciding what to do. Someone must be responsible to take corrective action and follow up on this action later.

This section will outline a system of control which has the ability to respond to change.

PREPARING A PRODUCTION ORDER

A production order is the document which authorizes production of a quantity of a product. While a production order may be produced by a computer, it must have the following features. It must serve as a vehicle for coordinating functions such as the issuance of material to various departments, starting production in more than one department, providing reference to standing instructions and specifications, and publishing special instructions. Issuing the order is a production control function.

Preparation of production orders occurs subsequent to the long term master schedule and prior to production of the short term schedule which is based on precise plant loading. Master schedules may be prepared for periods extending from two months to one year, but three months is considered a reasonable time period for apparel plants. Master schedules present definite production objectives, but can represent only tentative plant loadings. Plant loadings provide a more accurate match between work requirements and capacity for a short term than can be realized over a long term. The short term often used and recommended here is one to two weeks. Production orders may be written either before or after dealing with the loading and scheduling considerations. These are options. Generally, a production order follows the master schedule. Deviations may occur where piece goods are delayed or sales fail to develop.

In many apparel industry cases, the production order is prepared as stipulated by a master schedule. The quantity the order represents is then loaded against plant capacity to be in process for a normal production cycle (a week or several weeks).

Certain steps must be taken in order to load and schedule production orders to completion. They include:

a. Determining work requirements.

b. Determining the daily production rate.

c. Determining operator day-per-day requirements for a balanced work flow.

d. Determining work place requirements and availability.

e. Determining weekly style batches to schedule as part of the whole order.

f. Loading operations for balanced flow.

g. Preparing a daily schedule.

Because production order quantities do not necessarily conform to quantities which can be started and finished within one week, they often must be broken down into style batches for scheduling purposes. If all the scheduling analyses were to be completed before the production orders were prepared, it would be possible to prepare an individual production order for each weekly style batch.

Certain characteristics of production orders appear to be common to a broad range of manufacturing, although formats and technical features may vary. They identify the product which may include the style, model, family, product line and even customer. They specify quantities to produce component materials, the units of measuring them, the standard quantities per unit of product and the amounts to requisition for the order. The materials themselves must be clearly described.

Production orders should designate the plant which is to do the work. Orders should indicate which work centers or production lines within the plant will process the product. The listing of work centers or operation sequences is followed in creating work coupon tickets by which operators are paid.

Reference to both the bill of material and the operation sequence list should appear in the heading of a production order. Either document can then be quickly referenced or retrieved as needed. The production order is related in this way from the start to the fundamental sources of instructions.

Orders should bear the priority classification if one is assigned. Orders should carry the dates of preparation, issuance, scheduled start and schedule completion. Loading priority numbers provide a key for determining which orders are loaded as parts of the first group, a second group, etc. These would be grouped as order quantities going through process simultaneously.

Some batches may start and finish earlier or later than others. It is possible to establish a priority numbering system which provides for sequencing order starts within a major priority classification. Many plants color code the work tickets by date of entry in order to keep a cut together.

There should be and often is a place for special instructions or comments to appear on a production order. It is important to note if the service being provided the customer order is being expedited or if there are special customer delivery requirements. Failure to include special delivery requirements can lead to losses of customers, business as well as to inventory overstocks.

Currently most production orders are designed for batch processing. Recently unit processing planning is, in some cases, replacing batch processing. This change in procedure and controls will be reviewed separately.

Mass production has not existed in the apparel industry for many years, if it ever really did. Some people imagine it exists now but this is an illusion. Mass production is a process in which a very large or indefinite number of items of a homogeneous nature are being made continuously. There is a single route through a series of highly specialized operations. The operations are closely coupled and have very minimal inventory between them. Operators perform the same methods for a long time and become highly efficient. The organization of people and equipment is very inflexible and not adaptable to different products.

APPAREL PRODUCTION ORDERS

While production orders for apparel products are often called "cutting orders" or "cutting tickets," they usually contain much more information than just the quantities of piece goods and trim to be cut.

These orders authorize production of a specific product (or style) at a particular time, in a certain quantity, using particular materials, supplies, and equipment according to some special and standard instructions. They present marking and spreading plans for each textile material. They make the connection between all these elements. They may be produced manually or on a computer. Specific pieces to be cut may or may not be specified.

It is a production control function to issue a document which authorizes production and contains the basic information for the guidance of plant supervisors at each level of operations (i.e., cutting, sewing, pressing, packing, etc.). A production order for a quantity of apparel contains elements which may be classified under the following headings:

a. Identification.

b. Timing.

c. Summary of complete products ordered.

d. Cutting order section.

e. Trimmings and supplies specification.

f. Comments and instructions.

The identification section of the order should contain these items:

a. Plant.

b. Production line.

c. Work centers processing product (if unusual).

d. Style number and name (if any).

e. Model number and name (if used).

f. Style or model family.

g. Product line.

h. Customer or customers (if important to deliveries).

i. Bill of materials number.

j. Operation list number.

The part relevant to timing should include:

a. Date of order preparation.

b. Date of issuance.

c. Date to start work on order.

d. Date to complete whole order.

e. Priority number applicable between dates to start and finish.

A summary of the completed products ordered can be presented in the form of a matrix and can supply these items of information:

a. Sizes (across top of page).

b. Colors or designs (down left column).

c. Quantities at places where size and color correspond.

d. Total quantities of each color (down righthand column).

e. Total quantities of each size (at foot of each column).

Cut planning a production order involves both marker and spread planning as well as instructions for cutting and bundling. This activity involves some technical knowledge. Cut planning affects labor costs and material utilization. Many firms have developed computerized cut planning systems which are based upon "expert systems."

Usually this work may be delegated to a specialist within a production control department. Sometimes planning is passed along to the apparel plant's cutting department. In other cases, planning may be contracted out to a service center.

Marker and spread planning work is usually done manually in most apparel firms. Simple programs for using programmable calculators to plan cuts are in use. There are other programs for PC's and larger computers available.

Cutting orders should tell how each marker is made and how many plies are to be laid up for each different kind of fabric used. Markers can be described by section or by full table length. The following information should be specified:

a. Material.

b. Ordinal number of market (or any any ID).

c. Sizes to mark.

d. Assortment of sizes (size-multiples).

e. Color or design distinction of material.

f. Plies to lay up for each fabric.

g. Total ply height under the marker.

h. Total quantities of each size under the marker.

i. Total yardage of material (at standard usage) which should be provided for each color or fabric within each marker.

Because of the bulk, finish or other characteristics of each material, there may be different maximum ply heights for materials. This may result in a different number of markers needed by material. More than one material may be used in the style. The assortments of sizes appearing in single markers may be different from the overall assortment for the whole batch. The volume of the production order may require cutting a succession of markers. The central concern of production control is to make the final totals for each size and color in each material used add up to the total needed for the style without piece goods shortages or surpluses.

Some combinations of sizes on certain widths of materials provide for reduced usage of materials in many apparel styles. Styles using varied, large and small pattern shapes often have a characteristic which allows a planner to combine more sizes in the same marker to reduce yardage used per garment. Beyond a certain limit, there is normally no improvement. This factor should affect the number and kinds of markers made. Planning optimum markers is a way of reducing yardage consumption below standard usage. The planner is constrained by ordered quantities which may not permit optimization. Consulting firms offer computer programs which plan optimum markers.

There are two frequently encountered approaches for dealing with the provision of materials, trimmings, findings, and supplies in the apparel industry. The first is one in which a single production order is used as a combination of a materials requisition and authorization to produce. The second approach is one in which different documents are prepared for production authorization, marker and spread planning, and materials requisitions. These forms are specifically addressed to the concerned departments. Both approaches are in use. There may be circumstances in which one or the other approach has advantages. A typical example of a production order is shown as Form 1.

Authorization to produce normally proceeds to cut planning. This activity determines raw material requirements and posts inventory records (some firms do this by computer and obtain a printout showing roll number, length, width and location). Generally, this group also posts the records for other materials needed. Such things as labels, linings, and buttons are usually supplied by the box full or by the bundle, and so are over-supplied as a matter of course. Supervisors have the responsibility of returning surpluses to stock.

The quantities of trimmings, findings, and supplies which must be provided to the order should be calculated item by item. The item and standard usages are obtained from the Bill of Materials for the style. These standard usages per unit are exploded into total quantities of each item and recorded on either a production order or upon a materials requisition, depending on the approach being used. Standards usages establish the criteria by which a plant may be evaluated. Managers who meet or beat the standards may receive bonuses; those who overexpend standards may be investigated.

The special instructions and comments may tell anything relevant to producing the style. This part of the order may be used to:

a. Call attention to pieced parts in some markers.

FORM 1. APPAREL PRODUCTION ORDER

GROUP:	PLANT:	PROD. ORDER NO.:	
STYLE NO.:	STYLE NAME:	MODEL NO.:	MODEL NAME:
STYLE FAMILY	PRODUCT LINE:	CUSTOMER:	

WORK CENTERS DESIGNATED TO PROCESS ORDER:

DATE ORDER PREPARED	DATE TO START	DATE TO FINISH
DATE ORDER ISSUED	DATE INTO SEWING	PRIORITY NUMBER

SUMMARY OF COMPLETE PRODUCTS

COLORS OR DESIGNS	SIZES ORDERED														TOTAL
TOTALS:															

TRIMMING OR SUPPLY ITEM	UNIT OF MEAS	STD MEAS /EA	GARMS TO BE SUPPLD	QUAN TO SUPPLY	QUAN REALLY SUPPLD	QUAN REALLY RECOVD	QUAN USED /EA	QUAN LOST /EA

SPECIAL INSTRUCTIONS OR COMMENTS:

b. Require blocking and pinning of plaid materials.

c. Tell how to use certain findings (e.g. blue buttons on blue blouses, or white buttons on white).

d. Explain where to sew medallions.

e. Present any significant comment or instruction.

The ordinary practice in apparel making is to prepare the bundle list and the corresponding work coupon tickets for the order after the fabrics have been cut. The numbers on these documents must be exactly right. There is no tolerance for error.

There are a few firms that make up the bundle list and coupon tickets for production orders before they are cut. They require that all bundles be spread to the quantities specified, and that all fabrics be pulled properly. Other firms prepare these documents after cutting because it is expensive to duplicate and correct bundle lists and coupon tickets when cuttings have not conformed to the original order.

An example of an apparel production order suitable for a multi-material cutting of a style follows as Form 2.

FORM 2. MARKING AND SPREADING INSTRUCTIONS

STYLE:	PRODUC. ORDER NO.	SHEET_____ OF _____
MATERIAL:	WIDTH:	END USE:

MARKER NUMBER: COLOR OR DESIGN:														TOTALS
SIZES TO MARK														
SIZE MULTIPLE														
STD YDS/SZ MKD														XXXX
INCREM. LENGTH														
PLY STRATUM														
TIMES REPEATED														
PLY STRATUM														
TIMES REPEATED														
PLY STRATUM														
TIMES REPEATED														
TOTAL UNITS														
TOTAL YARDAGE														

MARKER NUMBER: COLOR OR DESIGN:														TOTALS
SIZES TO MARK														
SIZE MULTIPLE														
STD YDS/SZ MKD														XXXX
INCREM. LENGTH														
PLY STRATUM														
TIMES REPEATED														
PLY STRATUM														
TIMES REPEATED														
PLY STRATUM														
TIMES REPEATED														
TOTAL UNITS														
TOTAL YARDAGE														

MARKER NUMBER: COLOR OR DESIGN:														TOTALS
SIZES TO MARK														
SIZE MULTIPLE														
STD YDS/SZ MKD														XXXX
INCREM. LENGTH														
PLY STRATUM														
TIMES REPEATED														
PLY STRATUM														
TIMES REPEATED														
PLY STRATUM														
TIMES REPEATED														
TOTAL UNITS														
TOTAL YARDAGE														
GRAND TOTAL UNITS														
GRAND TOTAL YARDAGE														

10

Production and Cut Planning

INTRODUCTION

Pattern entry and grading computer activity usually occurs during the preseason planning cycle; however, the final entry and grading of patterns may be delayed until production is planned.

The objective of this section is to explain how the planning function develops specific directives for making markers. This chapter progresses from the general principles of planning production to the specific task of planning markers. Production planning and control sometimes do not consider the impact of sizing, marker length, etc., on marker planning.

The following sub-sections will cover this area:

a. Production planning principles relating to marker planning.

b. Building cutting orders.

c. Determining marker size assortments and spreading heights.

Production planning is concerned with developing long term plans and with implementing these plans within specific periods of time. A forecast or estimated demand for products is converted into an assignment of resources, human, equipment and material. Production planning consists of establishing a rate of production to meet sales requirements which minimize production costs. In the apparel industry, the first point of control over production rates normally is in marker planning.

ROLE OF PRODUCTION PLANNING

There are two principal roles which firms assume in production planning. First is a passive role in which a firm attempts to respond to demand by satisfying it through internal resources manipulation. Second is an active role which involves manipulating demand or creating demand. The firm makes an effort to influence outside demand.

Most firms, including apparel firms, adopt a mixed strategy in regard to these roles. However, most apparel firms are usually heavily involved in the passive role. They try to satisfy demand by changing the production related variables which they can control. They alter work schedules and vary the size of the work force in order to adjust production rates. This is the basic strategy of any labor intensive industry. Product mix is varied by adapting resources. Inventory levels are varied to match demand by altering production rates internally or by subcontracting work outside the firm.

Active roles may be played to some extent, but not to the same degree as the passive role. Firms may engage price cutting at times of lower demand or may offer discounts for volume purchases. Some firms, with well-known trademarks, involve themselves in seasonal advertising campaigns to promote styles as well as their trademarks. "Ship and Shore," "Playtex" and "London Fog" are examples of brands currently using this approach.

Information required for the production planning system will vary with the role the firm plays. However, these specific items of information are usually needed:

a. Economic conditions.

b. Market demand.

c. Raw material availability.

d. Knowledge about what competitors are doing.

e. Legal status or restrictions related to products (Example: fireproof pajama fabrics for children).

f. Size of current work force.

g. Inventory levels.

h. Production procedures, sequences and/or processes.

As production information is consolidated and utilized, a manufacturing schedule is formulated. This schedule becomes a time-table for performing activities, for utilizing resources and for allocating facilities. In order to schedule, it is necessary to specify ways to accomplish the job. It is necessary to specify:

a. Work places to support sequences of operations to be performed on products.

b. Equipment and personnel to specific work areas.

Segments of production capacity are then allocated to serve specific production requirements. Incoming sales should be compiled in such a way that practical cutting orders can be written. Job priorities must be established. Cutting orders are prepared for specific production time periods. Orders must be issued on time, to initiate the jobs.

Production planning and control people usually are not knowledgeable about material utilization techniques. Their duties do not bring them in contact with such considerations. There is normally no real conflict of interest of production planners with manufacturing efforts to structure markers to improve utilization. However, they may think that their production orders are being changed when markers are restructured to lower piece goods costs.

Managers of marker making and cutting operations are the people who have the most interest in economic material utilization. Good utilization may promote personal advancement and job retention. People making markers on a computer are often forced by the pressure of circumstances to give first importance to making markers on time. As a result, they tend to keep a log of markers made to use when under pressure. Some marker systems automatically store old markers and rank them by efficiency.

Normally, shorter markers can be produced faster. Both marker makers and production planners usually prefer shorter markers. Shorter markers are less likely to allow pattern arrangements which achieve the best material utilization, that is, the least yardage per garment.

People who can communicate can adjust activity between production planning, marker planning and cutting. For example, a marker could combine a series of small production orders, in one style, to improve material utilization. After cutting, the small orders can be separated and can be fed into production in a timed sequence acceptable to production planning. The result could be a savings in both labor and material.

Another area of compromise is the matter of cutting to exact quantities. If the firm's policy allows a few additional sizes to be cut occasionally, it may be possible to produce fewer, longer and less costly markers. Over-cutting selected sizes does not normally create any serious, practical problems. Actually, cutting exact quantities can be a worse problem. During sewing, some garments may be damaged. Exact cutting could then cause delivery of short quantities.

Ultimately, production planning must have the last word upon what shall be produced. The production planning department receives sales data, and they allocate capacity to meet delivery dates. Materials availability is tracked by the production planners.

Marker making is the point at which the scheduling is first implemented. The importance of making timely markers is critical to customer service. Time compression between issuance of cutting orders, the delivery of cut work to sewing, and finished goods to shipping is made more acute as the number of cycles multiply. Pattern grading and marker making, by traditional methods, have become too slow. Computerized marking systems allow apparel firms to increase the number of styles in production. These computer marking systems satisfy the need for swift grading and marking.

People who buy apparel give no indication of any desire for standardized dress. This suggests that the conditions which have stimulated the development of the computer marking systems will persist and intensify so as to make the use of such systems indispensable in the future.

PLANNING MARKERS

Marker planning consists of selecting from a cutting order sizes to mark and establishing ply heights for markers in order to satisfy a complete cutting order. The production, planning and control organization normally will have dealt with the problems as to style quantities to be produced, allocation of production capacity needed for styles running concurrently, and as to whether supplies and materials are available. The scope of marker planning is usually limited to the immediate practical problem of establishing appropriate sizes and spreading heights needed to achieve the production cutting order.

Styles or patterns are usually predefined within the marking system (manual or computerized). They must be identified on the cutting order. Sizes to be cut are named. Garment quantities to be produced by size are given. Normally a maximum or a desirable ply height is stated. The maximum height to which the cloth can be stacked up (or piled) for cutting normally has been predetermined for each type of material. Piece goods to be cut are identified and their lengths and widths should be given. These activities are the true planning tasks. Marker planning work is defined here as something distinctly different from making markers.

Marker making consists of arranging patterns within a specific rectangular area so as to occupy an area with the least lost space or waste of material. This requires a person with a high order of geometric judgment and good perception of spacial relationships between pattern shapes. (A jigsaw puzzle addict does well.) Experience with certain pattern sets becomes becomes valuable to those people who arrange patterns in markers whether on a computer tube or on a table. Experience makes it possible to find the more effective arrangements in less time. Some computers have algorithms which make a marker. The marker maker starts with the computer solution and improves upon it.

When people speak of "optimizing" markers, the same distinction must be continued between the planning and the doing. Optimization, with respect to planning markers, refers to finding the best combination of sizes, garments marked per size, with the material width and ply height, to enable the most economical cutting, considering both material and labor cost.

Optimization in making markers involves finding the best arrangement of pattern shapes to shorten the rectangular area of a marker within the fabric width, the style constraints and the assortment of sizes. Marker optimization tries to achieve the best results from the constraints established by production planning.

Computer programs have been developed to provide mathematical optimization techniques which work well in both the planning and the marking areas. So far, it has been found that there is a significant variety of approaches in both areas which end with approximately the same results. Lectra, Microdynamics and Gerber offer programs to assist in marker making.

Techniques used to optimize marker heights include:

a. Simplex method of linear programming.

b. Nonlinear optimization (for example, the method of steepest descent).

c. Dynamic programming.

d. Combinatorial optimization.

Currently, mathematical optimization is not universal for all apparel products. Therefore, this discussion of marker planning will deal with the traditional application of logic and arithmetic. Traditional approaches have, more often than not, produced better results in marker planning and making than very sophisticated mathematical techniques.

Each kind of fabric which might go into a garment might have some different characteristics. One of the most significant of these to the planning of markers is the bulk of the material. Bulk causes the number of plies possible to spread to differ from fabric to fabric. Fabric weight also alters the bundle size that a sewing operator can handle comfortably. This affects the cutting height and how bundles may be split out of a cut.

When the cloth which goes into a garment includes diverse materials for shell and lining materials, the maximum ply heights possible may be different for the different fabrics used in a single style. The maximum ply height should be specified as a characteristic of each type of material. (Normally cutters make tests to establish these limits.)

Many garment manufacturers use a standard number of plies in a bundle. Bundle size is related to weight, bulk and pay systems. This means that the spread is laid up in standard numbers of plies separated by a paper or by color. Bundle size may vary from firm to firm and from one fabric density to another. Some plants may use a standard 24 ply; others may use 36 or 48. The standard bundle size is ordinarily used to determine the maximum practical ply height of a spread in terms of multiples of a standard bundle.

It is necessary to know the number of component parts to make a complete garment. Some computer marker systems have limitations on the number of patterns per marker. (This limit is established by the size of the computer memory and by the vendor's method of describing apparel garment parts.) The number of garments which it is possible to include in any given marker depends upon the number of parts in the style. The garments per marker limitation may be computed by dividing the computer or the table length limit by the number of garment parts per garment, whichever is less.

Once the marker limits have been established, marker planning can proceed through a series of routine steps. In some cases, there may be complications or alternative approaches which lengthen the routine. Most production organizations use standardized procedures and some shortcuts, of one sort or another, to do marker planning. They do this to save time. Standardization makes it possible for people with lower spatial configuration qualifications to make markers. Most cutting operating personnel are not accustomed to seeing the whole process. They are often astonished to learn how complex the marker planning job is.

Marker planning deals with one cutting order at a time. This usually involves one style and may involve several different markers in each of several materials. The steps to take in planning markers fall under the following general headings.

a. Obtain a cutting order which furnishes both the authorization to make markers and the description of the purpose.

   Identifying the:

   (1) Style.

   (2) Sizes ordered.

   (3) Quantities to cut in each size.

   (4) Quantities to cut in each color or design.

b. Segregate and classify the markers which must be made according to:

   (1) Material.

   (2) Width.

   (3) Style.

   (4) Component parts of the garment in each material.

c. Collect and analyze data relevant to the limitations applicable within each classification of material:

   (1) Maximum number of patterns per marker.

(a) Table length limitation (Old times).

(b) Computer program system limitation (New times).

(2) Component parts per garment, per type of material.

(3) Maximum number of garments per marker.

(4) Standard number of bundle plies per material.

(5) Maximum possible plies in a spreading height.

(6) Maximum practical plies in a spreading height.

(7) Maximum practical height in standard bundles.

(8) Sizes ordered.

(9) Average yardage (or metric length) of material per garment, either taken from experience or estimate.

d. Describe the markers to be made in each material and width, determining:

(1) Ply height.

(2) Sizes, to include in each.

(3) Quantities to cut in each size.

(4) The assortment of sizes in the marker.

e. Document the results of planning markers as:

(1) Instructions to marker maker.

(2) Record of markers ordered made.

(3) A statistical history of styles.

(a) Ordered assortment of sizes.

(b) Width cut.

(c) Material utilization achieved.

Certain general aims are sought in marker making. Producing markers in a hurry to satisfy a production need conflicts with the objective of securing the best material utilization.

The basic objective inherent in marker planning is primarily concerned with securing the best utilization of material. Wherever a choice between saving material, rather than labor, is made, material is normally favored. However, many firms calculate material costs versus labor savings on a computer in order to select the least-cost combination.

As a greater proportion of a firm's output is cut using automatic cutting equipment, the labor cost of cutting additional garments is relatively less important. Material savings gain importance. Manual cutting costs remain relatively greater. Therefore, extra garments will usually not be cut unless they fit into an existing pattern of cutting.

In planning markers it has been found that when large and small sizes are combined evenly, piece goods are usually saved. Combining the larger and smaller sizes provides a greater variety of pattern shapes to fit together compactly.

Marker planning usually tries to combine the largest acceptable number of garment sizes within the table length or computer limits to make a marker. Replanning markers (iterating size combinations) as often as possible to save piece goods will sometimes result in markers having less than the maximum ply height, yet being more efficient. In this case calculation of the additional labor should be made.

Material utilization statistics must be collected to support planning decisions. Data must be accumulated from experience by pattern set or style, by width of material, by class of pattern arrangement, and by garments per marker. Pattern arrangement is often controlled by the spreading method required, for example, face-to-face two ways, face-up one way, face-up two ways, or face-to-face one way. Sometimes further definition is required. These may specify pattern orientation and restrict pattern rotation or pattern flipping.

There are widths of material which accommodate combinations of patterns of some styles well while other widths create piece goods losses. Data has been assembled by some firms on the impact of width on various styles. A standard group of sizes is used to calculate this impact. Cutters then advise piece goods buyers which widths to buy. This data is usually presented as a wavy, oscillating curve, reflecting the fact that there are more favorable and less favorable widths.

While acknowledging that variations do occur in the results, the following general rules offer the best probability of success at utilizing material:

a. Combine as many garments into a marker as circumstances permit.

b. Combine large and small sizes in pairs in the same marker.

c. Combine sizes near the middle of the range in the same marker.

d. Try to obtain fabrics in favorable widths for marking, if possible.

e. Examine patterns to determine how contact points may be changed in order to make boundaries approach each other more closely. (Cut off all seamage corners feasible to cut.)

Longer spreads save spreading labor costs over laying up short spreads, particularly when high speed, computer controlled, large roll spreaders are used. Currently the mapping of material flaws is under test. The textile firm marks the selvage with magnetic dots to locate flaws. The spreading machine reads the selvage for dots and halts as flaws occur. The spreader then cuts out flaws. Some spreading machines are testing automatic flaw cutting. In the experiments the spreading machine contains marker information to guide it.

Cutting high spreads reduces labor costs below cutting low spreads. More units per knife cut are produced. Better material utilization practices in marker planning will tend to create longer spreads, but sometimes may result in less than maximum spreading height. Lower spreads are not as significant to cutting costs for those who use computer cutting systems. All present computer cutting systems cut fewer plies than manual knife systems.

MARKER OBJECTIVES

One of the first objectives in planning markers for a cutting order is to define the scope of the job. This includes: computing the maximum practical ply height of the spreads, summing up the number of garments in the order, and computing the number of garments which theoretically could be put into a marker when spread to the maximum height.

A normal first step in planning is to calculate the maximum practical ply height. The maximum possible ply height for the fabric is divided by the standard bundle size. The even quantity resulting is saved and the fraction is discarded. This establishes the final ply height (i.e., six bundles of 30 equal 180 plies). The result is the maximum practical number of plies to be laid up in a spread of that cloth. (All of the above hand/knife cutting calculations are abrogated by computer cutting limitations.)

The number of markers which could be made from the cutting order, to be spread full height, can be computed by dividing the foregoing total number of garments spread full height by the maximum number of garments possible per marker. Only the integer part of this dividend is interesting. If this dividend amounts to less than 2, then it will probably be worthwhile to examine possibilities other than spreading to maximum height. If the result is greater than 2, then at least one and maybe two markers can be planned to cover spreads going into full height. The fractional parts in the foregoing numbers of markers simply indicate that there will be short markers covering lower spreads upon which some of the remnant pieces may be laid. (Note: There are computer programs which calculate marker heights. If you know the rules one can easily write special programs by product and fabric.)

At this point, the number of full height markers that can be laid up for the cutting order is known. Also, the sizes in the range, the quantities of each size and the maximum number of garments which can be put into one marker are all known. The assortments in these markers have not yet been determined. There are various ways to do so. The method described below is a rather mechanical procedure chosen to serve the following ends:

a. Provides markers with size quantities which do not exceed the ordered quantities.

b. Pairs up large and small sizes from the extremities of the size scale proceeding toward the middle.

c. It is mechanical enough to be readily programmable on almost any computer.

The assortments of these markers, to be spread to maximum practical height, should be composed combining large and small sizes, alternating with opposite ends of the size scale in the following order:

a. The smallest.

b. The largest.

c. The size next to the smallest.

d. The size next to the largest.

e. Continuing on in this same manner of alternation until one of two things happens:

    (1) The maximum number of garments which can be put into a single marker is reached; in this case, stop.

    (2) The opposite ends of the size scale are reached in the continuous sequence of alternation without exceeding the maximum number of garments per marker; in this case reverse the direction of alternation and proceed toward the opposite ends of the size scale again, until the maximum number of garments is reached.

This scheme of altering size combinations tends to select more sizes near the center of the distribution of quantities over the size range than at the extremities of it. The reason for this is that each opposite side combination subtracts a number of garments from the size quantity equal to the maximum practical height of the spread, and it does not subtract from quantities less than the maximum practical height. As a result, this procedure builds the larger part of the assortment where the larger part of the size quantities are. The remainders under each size are left to be dealt with, using spreads going up to less than maximum height. Altering of size combinations will provide compatible equal quantities of sizes as the method reaches from the end sizes through the mid-points of the size range. This activity continues until the maximum height can no longer be subtracted from the production order. This finally leaves a positive remainder which normally is planned in supplementary end lays.

The alternative size selection procedure builds paired-up size assortments and produces even numbers of garments per marker. It continues until the limiting number of garments per marker is reached or until the size quantities begin to be exhausted, whichever occurs first. (Note: There are some apparel products that combine best in units of odd numbers such as 3,5,7,9,11, etc. However, the procedure of blending large and small sizes still works best in reducing piece goods loss in this marker situation.)

Determining the size assortment and ply height for a marker combining largest feasible number of sizes at less than the maximum practical ply height involves a different procedure. If only a few of the sizes in a production order have quantities that are spreadable at the maximum height, then in order to combine other sizes in the marker, the spread cannot be laid up to maximum height and still achieve the size pairing combination piece goods savings. The size assortment must then be adjusted to permit a larger combination of sizes at a lesser spreading height under one or several markers. It is not unusual to find production cutting orders where it is not profitable or possible to plan the markers at the maximum practical ply height. In these cases, it is better to combine a larger assortment of sizes at a lower ply height so that piece goods savings may be achieved.

This type of marker planning consists of finding numerical values for lesser size quantities which divide the cutting order size quantities into acceptable cutting heights. Cutting heights should be composed of standard bundle sizes. This planning activity can be achieved by converting the size quantities into whole numbers of standard bundles before establishing a marker and by disregarding remainders. Remainders can be cut separately.

PLY HEIGHTS

Not all cutting tickets will contain sufficient garments to create markers containing the maximum number of garments and/or the maximum practical ply height. Markers may be planned as follows:

a. Convert each size quantity into an integer number of standard bundles.

b. Rank these numbers of standard bundles in descending order.

c. Sum the number of bundles for the order.

d. Divide the smallest number in the ranking into the sum of bundles per order.

e. Compare this dividend to twice the maximum possible number of garments per marker. (This is because two main markers are being planned.)

   (1) If the dividend is greater than twice the maximum number of garments per marker, go to the next larger number of bundles in the ranking.

   (2) If the dividend is less than or equal to two times the maximum number of garments per marker, use this size quantity in bundles to perform the next division. (See f below.)

f. Divide the largest quantity in the ranking by the size quantity of bundles, not zero, which produces a number which must be made an integer, if it does not occur as one. (Note that fractions reduce to integer zero.)

g. Multiply the foregoing integer by the standard bundle plies to determine the ply height of the spread.

h. Determine the assortment of sizes by dividing each size quantity by the ply height and record the integer multiple under each size, disregarding fractions.

i. Separate the remainders, per size, as follows:

   (1) Multiply the ply height by the integer multiple in the assortment.

   (2) Subtract the product from the total size quantity.

   (3) Record the remainder under the size if it is positive and not zero.

j. Divide the assortment under each size into two parts as follows:

   (1) Divide the multiple by two.

   (2) Make the dividend an integer and record it under the size.

   (3) Subtract the foregoing integer from the whole original value and record this difference under the size.

   (4) Do this for every size and record the divisions at the separate assortments of the two markers having the same ply height.

Odd numbers in the assortment of sizes may be broken into two parts per size using this last tactic. The result is to make one marker contain more garments than the other. Since the two assortments together do not exceed two times the maximum number of garments per marker, it is unlikely that separating them so near their mid-points would cause one of them to contain more than the maximum number of garments. It could happen when two markers contain nearly the maximum number of garments, and, if it did, would require that a garment be switched from one to the other.

MANUAL EXAMPLES OF PLY HEIGHT CALCULATION

Two examples follow. The first case deals with a large cutting order which contains quantities that justify making several long markers for spreads at maximum height. The second case covers the smaller cutting order, having no more than two or three major markers, plus short ones for remainders. The major markers probably will contain less than the maximum number of garments. They may either be laid up to heights less than the maximum or they may contain fewer than the maximum number of garments per marker. From the technical standpoint, the lesser cutting order may occasionally present a greater problem than the larger. Nevertheless, it can be solved methodically.

The following examples use boys' long sleeved sport shirts as the subject product. These have a moderately large number of parts per garment, all in one material. Having only one fabric makes the presentation of examples considerably shorter. All the work that will be presented must always be repeated for each different kind of material going into the same garment. This boys' sport shirt will have two fronts, no pockets, two collars, two yokes, two sleeves, two sleeve slip binding strips, four cuffs, and one back, which make a total of fifteen parts per garment.

When the following data pertains to the cutting order for example:

Maximum total of patterns per marker	200
Count of sizes in the order	5
Maximum possible number of plies in height	329
Standard number of plies per bundle	48
Component parts per garment	15

The cutting order comprises these:

Sizes/Quantities:

$$\frac{10}{2000}\ \frac{12}{4160}\ \frac{14}{6132}\ \frac{16}{4052}\ \frac{18}{2064} = \begin{array}{c}18{,}408\\ \text{Total Units}\end{array}$$

It is possible to deduce from the given data, the following facts:

Maximum practical ply height	288
Practical spreading height in standard bundles	6
Maximum number of garments per marker	13

Number of markers at maximum size	3 (4 possible)
At less than maximum	2 (1 possible)
Maximum number of garments per spread	3,744

There could be as many as four spreads at maximum size and one at less. In order to maintain the greatest possibility of pattern combinations, the east two spreads will be put together for planning purposes, in order to make both markers contain as many garments as possible. The purpose in doing this is to provide the best opportunity for interlocking pattern shapes in such a way as to make the best use of the area of the material within the width of the cloth.

Now the alternating size selection method will be employed to build marker assortments, within the limiting number of garments per marker, for three markers:

Sizes/Quantities:

10	12	14	16	18		18,408
2000	4060	6032	4052	2064	=	Total Units

There will be two more major markers needed following these three at maximum size to complete the whole order. These spreads will not be quite so large in some respect. Either the ply height or the number of garments per marker or both will not be required to be the greatest possible. The remainders left in various sizes will be small enough that they may be completed on short markers, perhaps using up remnant pieces of cloth.

The alternating size selection method to build the biggest assortments of sizes is illustrated on the following page.

Beginning at the low end of the ranking, select and try the values, one after the other, going toward the high end until a divisor is found for the total number of bundles that will give a number less than twice the maximum number of garments per marker. It will go like this:

$$(1)\quad \frac{147}{3} = 29.4 > (2 \times 13)$$

$$(2)\quad \frac{147}{7} = 21.0 < (2 \times 13)$$

The ply height would be 7x48 = 336 plies. However, 336 plies > the 329 plies of absolute maximum height stated at the beginning of this, and also 336 > the 288 plies of maximum practical ply height which is the greatest height to which standard bundles of 48 can be laid underneath the 329 plies of absolute maximum height.

Since the seven bundle height is too great, one bundle is removed and the comparison with the maximum practical ply height is made again. This time it is a usable height.

$$(7 \times 48) - 48 = 288 \text{ plies}$$

$$(7 - 1)(48) = 288 \text{ plies} = 6 \text{ bundles}$$

First marker with 13 garments, covering 288 ply, using digits:

Sizes:		10	12	14	16	18		
	(1)	288				288		
	(2)		288		288			
	(3)			288				
	(4)			288				
	(5)		288		288			
	(6)	288				288		
	(7)		288		288			
	(8)			288				
Totals:		576	864	864	864	576	=	3,744 Units
Assortments:		2	3	3	3	2	=	13 Garments
Balance:		1424	3296	5268	3188	1488	=	14,654

Sizes:	10	12	14	16	18		
Balance:	1424	3296	5268	3188	1488	=	14,654

Second marker with 13 garments, covering 288 ply:

Sizes:		10	12	14	16	18		
	(1)	288				288		
	(2)		288		288			
	(3)			288				
	(4)			288				
	(5)		288		288			
	(6)	288				288		
	(7)		288		288			
	(8)			288				
Totals:		576	864	864	864	576	=	3,744 Units
Assortments:		2	3	3	3	2	=	13 Garments

Sizes:	10	12	14	16	18		
Balance:	848	2432	4404	2324	912	=	10,920

Third marker with 13 garments covering 288 ply:

Sizes:	10	12	14	16	18		
Totals:	576	864	864	864	576	=	3,744 Units
Assortments:	2	3	3	3	2	=	13 Garments
Balance:	272	1568	3540	1460	336	=	3,176 Units
1/288 Ply:	0.94	5.44	12.29	5.70	1.17	=	29.92 Multiples

Convert the balance into standard bundles:

Sizes:	10	12	14	16	18		
Bundles:	5	32	73	30	7	=	147 Std. Bundles
Remainder:	32	32	36	20	0	=	120 Units

Rank the values of standard bundles per size:

Sizes:	10	12	14	16	18		
Bundles:	73	32	30	7	5	=	147 Std. Bundles

The ply height of the last two markers has been determined to be six bundles or 288 plies. The assortments will be developed on this basis and then the remainders will be determined. The best ply height for the last two markers may sometimes be less than the maximum practical height but this time the maximum is found to be best.

The last two major marker assortments are derived as shown below. The balance which was converted into standard bundles is used. The remainders, previously found, are ignored for the moment. The marker height will be six bundles.

Sizes:	10	12	14
Bundle Balance:	5	32	73
Assortments:	-	5	12
Quantities to Cut:	0	1440	3456
Preceding Balance:	272	1568	3540
Remainders:	272	128	84

16	18			
30	7	=	147	Std. Bundles
5	1	=	23	Garments
1440	288	=	6,624	Units
1460	336	=	7,176	Units
20	48	=	332	Units

The assortment of sizes above which contain 23 garments must be subsequently separated into two parts containing less than 13 garments each so that they may be made into markers, using the computer marker making system. This is done as shown here:

Sizes:	12	14	16	18		
Assortments:	5	12	5	1		
Half:	2.5	6	2.5	0.5		
Integer:	2	6	2	0	=	10
Difference:	3	6	3	1	=	13

Two markers can be defined using these assortments. The following restates them:

Sizes:	12	14	16		= 288 Plies
Assortments:	2	6	2		= 10 Garments
Sizes:	12	14	16	18	= 288 Plies
Assortments:	3	6	3	1	= 13 Garments

This concludes the planning of five major markers for a large cutting order, developing the ply heights and the assortments for each one. It should be observed that this task was accomplished by means of a routine, logical and numerical procedure after certain general decisions were stated in advance for planning the marker.

Markers for the remainders above would be provided additionally as follows:

Sizes:	10	12	14	16	18	
Quantities:	272	128	84	20	48	= 552 Units
Std. Bundles:	5	2	1	0	1	= 9 Bundles
Remnants:	32	32	36	20	0	= 120 Units

First Remainder Marker:

10	12	14	18	= 48 Plies
5	2	1	1	= 9 Garments

Second Remainder Marker:

10	12	14	= 32 Plies
32	32	32	= 3 Garments

Remnant Markers Completing Remainders of Orders:

14	= 4 Plies
1	= 1 Garment
16	= 20 Plies
1	= 1 Garment

An example of a smaller cutting follows. This, also, will be a boys' long sleeved sport shirt without pockets, having 15 parts per garment. The order will call for a quantity less than would be produced by two markers with the maximum number of garments covering spreads of the maximum practical height and greater than that which would be produced by one such marker in a spread.

The same initial data pertains to this cutting order as did to that of the first example:

Maximum number of patterns per marker----------------------200

Count of sizes in the order-------5

Maximum possible numbers of plies of height-------------329

Standard number of plies per bundle-----------------48

The component parts per garment--15

It is possible to deduce these items of data from that given:

Maximum practical ply height----288

Practical spreading height in standard bundles-----------------6

Maximum number of garments per marker---------------------------13

Maximum number of garments per spread-----------------------3,744

This second cutting order includes the following quantities per size:

Sizes:	10	12	14
Quantities:	686	1416	1835
1/288 Ply:	2.38	4.92	6.37
Integers:	2	4	6
Total Standard Bundles:	14	29	38
Std. Bundles in Markers:	12	24	37
Remnant Standard Bundles:	2	5	2
Remainder Units:	14	24	11

16	18			
1200	853	=	5990	Units
4.17	2.96	=	20.80	Multiples
4	2	=	18	Garments
25	17	=	123	Bundles
24	12	=	108	Bundles
1	5	=	15	Bundles
9	37	=	86	Units

Dividing through the size quantities using the maximum ply height, multiples of height are obtained which add up to less than the 26 garments that would be found if there were garments enough to justify two spreads and two markers of maximum size (2 x 13 = 26). Observation of the integer multiples of the assortment for a spread of maximum practical ply height reveals a divisible integer assortment of 18 garments which can be broken into two parts of nine garments each, which are either one less than the maximum of 13 garments per marker. These two markers leave remainders in whole standard bundles and in units as well. For this reason, it will be necessary to consider making a third major marker as well.

The assortment to be broken apart, and the parts themselves appear like this:

Overall Assortment:

	10	12	14	16	18		
Sizes:	10	12	14	16	18	=	288 plies
Assortments:	2	4	6	4	2		18 garments
First Part:	10	12	14	16	18	=	288 plies
	1	2	3	2	1	=	9 garments
2nd Part:	10	12	14	16	18	=	288 plies
	1	2	3	2	1		9 garments

After the foregoing markers are planned, the following remainders appear on each size:

Sizes:	10	12	14	16	18		
Remainder:	110	264	107	48	277	=	806 g
Std. Bundles:	2	5	2	1	5	=	15
Remainder:	14	24	11	0	37		

Rank the standard bundles in descending order:

Sizes: $\frac{18}{5}$ $\frac{12}{5}$ $\frac{14}{2}$ $\frac{10}{2}$ $\frac{16}{1}$ = 15 Garments

Dividing the 15 bundles by 1 yields more than the 3 garment maximum per marker. One standard bundle cannot be the ply height of a single marker in this case. Going to the next larger value, 2, divide the 15 by it to get a number less than the 13. It would be possible to make a marker with a ply height equal to two standard bundles. This marker would be described as shown:

Sizes:	10	12	14	16	18	=	96 Plies
Assortments:	1	2	1	0	2	=	6 Garments
Quantities:	96	192	96	0	192	=	576 Units
Remainders:	14	72	11	48	85	=	230 Units

The remainders would be made from remnant pieces laid on short markers. Completing this order requires three major markers and the usual odd single and double garment markers to use up remnants and make remainders. In summary, these are:

Two of these:

$$\frac{10}{1}\quad\frac{12}{2}\quad\frac{14}{3}\quad\frac{16}{2}\quad\frac{18}{1}\quad \begin{matrix} = 288 \text{ Plies} \\ = 9 \text{ Garments} \end{matrix}$$

One like this:

$$\frac{10}{1}\quad\frac{12}{2}\quad\frac{14}{1}\quad\frac{18}{2}\quad \begin{matrix} = 96 \text{ Plies} \\ = 6 \text{ Garments} \end{matrix}$$

Singles: $\frac{10}{1} = 14$ Plies

$\frac{14}{1} = 11$ Plies

Doubles: $\frac{16}{1} = 48$ Plies

$\frac{12}{3} = 36$ Plies

$\frac{18}{2} = 43$ Plies

These examples should illustrate most of the problems and tactics to deal with them involved in determining spreading ply heights and marker assortments. There are variations in the way things may occur which are too numerous to treat individually. For example, certain assortments may be required to be cut repeatedly in a large cutting order to serve a customer's delivery requirements.

In such case, those required assortments would be used instead of applying the alternation procedure described earlier to build an assortment. Another thing may influence assortments too. If the authority issuing the cutting order would permit the cutting of a few additional garments occasionally, some garments which would be cut uneconomically under short markers might be included in major marker assortment where better material utilization could be realized.

A shortcut often used is the combining of a few standard assortments. When the ordered quantities are adapted to the assortments available instead of the opposite being the case, the situation is made much simpler. For example, take the following four assortments of four garments each:

	S	M	L	XL
1	1	2	1	
2		2	2	
3		1	2	1
4	1	1	1	1

It is possible to combine these markers for use on single spreads, obtaining various different assortments. Observe:

	S	M	L	XL
1 & 2	1	4	3	
1 & 3	1	3	3	1
1 & 4	2	3	2	1
2 & 3		3	4	1
2 & 4	1	3	3	1
3 & 4	1	2	3	2

If the foregoing six assortments of sizes in addition to the originals do not cover all needs, then markers may be combined again. For example, marker number 4 may be combined again with each of the foregoing to create the following assortments:

	S	M	L	XL
1 & 2 & 4	2	5	4	1
1 & 3 & 4	2	4	4	2
1 & 4 & 4	3	4	3	2
2 & 3 & 4	1	4	5	2
2 & 4 & 4	2	4	4	2
3 & 4 & 4	2	3	4	3

When an order is received, an assortment of sizes can be determined from the size quantities and then the nearest approximation that can be made to it by combining standard assortments is used.

The arithmetic involved here is exceedingly simple. The procedure has the advantage that almost anybody can use it.

The disadvantage in combining markers is that short markers are always used. Even when laid end to end making a long spread, the markers determine a yardage per garment consumption which is always associated with a four garment combination, and never better.

The marker making effort can be substantially reduced by duplicating a stock of four standard markers for a size. Even when the assortments do not always serve well, some firms have persisted in using this procedure anyway simply in order to save marking time. This is an old fashioned practice which is being resurrected in somewhat different form by the computer marking systems. Computer systems can organize and store large numbers of markers for reuse. Computer programming is already available from several sources for marker selection purposes.

Any computer marker selection program is limited to the marker assortments which happen to be stored. Usually the stored markers are arrayed in sequence of material utilization efficiency.

There are situations where combining old markers is the best and the quickest thing to do. There are other cases where it is best to plan and make new markers. Marker makers should be able to evaluate the best way to do the job.

11

Plant Loading and Cost Control

INTRODUCTION

In the preceding sections, plant loading as performed by the Production Control Department has been discussed. Basically, it was stated that the production control staff prepares a work order or a cutting ticket to initiate production in those plants which have computerized payroll systems. The cutting order is converted into:

a. Bundle lists.

b. Work coupon tickets.

c. Routing sequences, if required.

These items in turn are used to prepare:

a. Piecework payrolls.

b. Work in process reports.

c. Operator performance reports.

d. Finished goods to inventory.

e. Excess cost reports.

These reports are a by-product of a computerized system. They can be arranged in many ways according to the user's needs.

PREPARING BUNDLE LISTS

Apparel products when cut may contain one or several plies. Normally, the plies will be divided into bundles. Bundles may contain a standard number, such as 24, for payroll calculation or for weight control.

Bundles for any one size and color will be uniquely numbered. Every part which goes into that one size and color will have the same serial number. Numbering is necessary for identification and control whether there are many bundles of different parts or one bundle with all the parts of the item. The rules for bundle size (number of plies) and for payroll ticket calculation are put into the computer and are revised as circumstances require.

Computerized bundle tickets are serially numbered for unique identification. They are listed for each production order. All plies within a bundle must be kept together to maintain shade control during the manufacturing. Work locations are maintained by tracking the location of a bundle between operations. The completion of work by an operator is counted as the completion of a bundle.

USE OF DATA BASE TO LOAD AND SCHEDULE PLANT

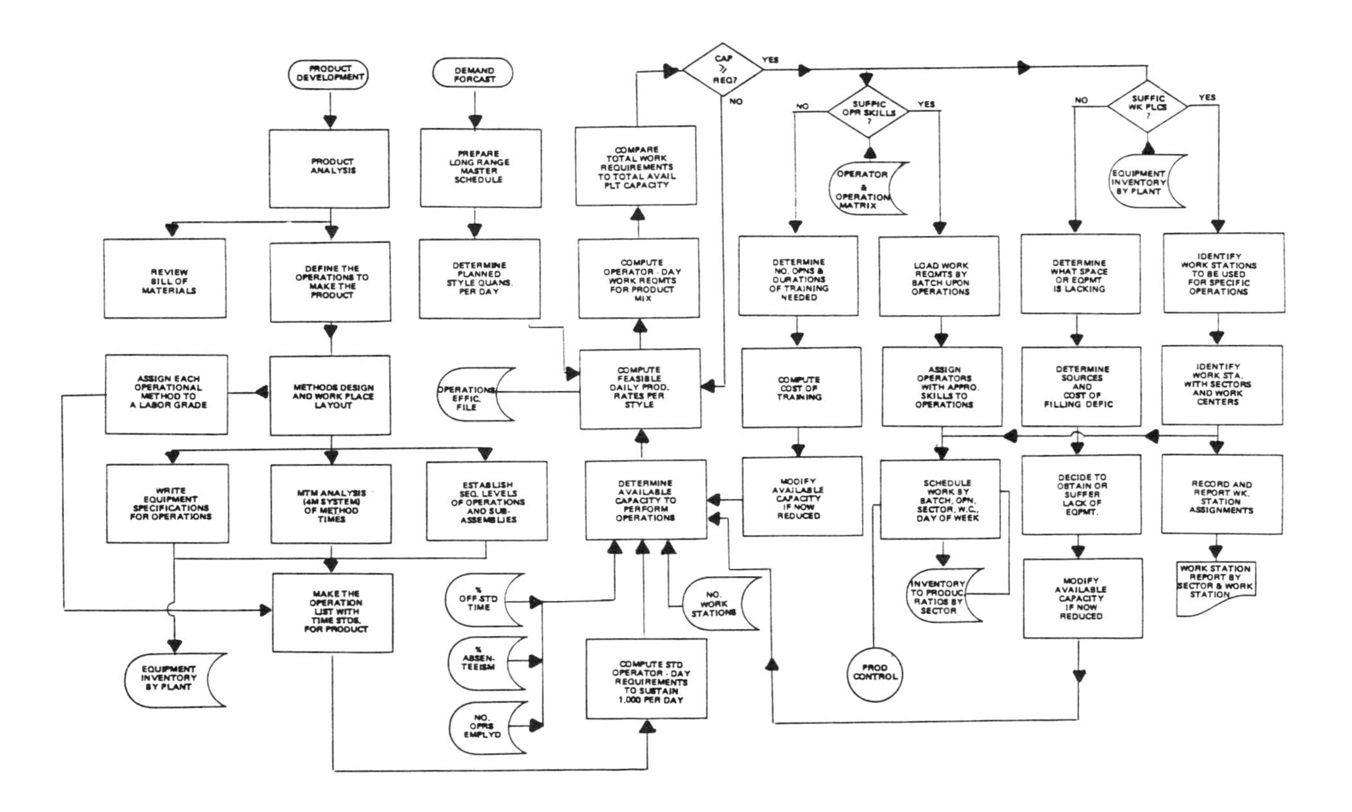

Before any portion of a production batch leaves a cutting department or any first sector of production, a bundle list should be prepared, itemizing each bundle. Normally, this is done by the computer. The following items of information must be checked for each bundle:

a. Serial number.

b. Size.

c. Color.

d. Quantity.

The bundle list itself pertains only to the individual production order. Therefore, the model or style is already established. When there is only one color per style and the production order provides that several colors are cut together under the same model number, it is still probably better, certainly more explicit, to state the size and color than simply to use the style number in place of these.

COUPON TICKETS

After the bundle list is prepared for end production order, computerized systems may make optically scannable coupon tickets for attaching to the bundles. There should be a coupon for each operation performed on the parts in the bundle.

Preparing the coupon tickets involves grouping the operations appearing on the operation list. These groups are maintained for various styles in the computer by the plant engineer. Assembly route codes, sequence level numbers and the designation of the next item following the current one are provided in the operation file. The operation file provides the references, the same every time.

The operation list for the style, the bundle list for the production order and the format of the coupon ticket are all stored in memory for computerized manipulation. Computers print the serially numbered coupon tickets for attachment to bundles.

Even though coupon tickets are organized into the same format for the same style, this does not preclude using alternative work routings. At any sequence level, there may be, and often are, alternative operations which can be started. This change depends upon which operators are free. The listing of coupons does not require that operations be performed in the listed sequence if there are other sequences possible and convenient. The important element is to keep operators busy and to keep work flowing.

When an operator completes work on a bundle, normally the coupon is torn off the bundle ticket and is stuck on a gummed report sheet. (Note: Some systems provide for coupon scanning at the work station. This action does not require using a gummed work sheet.) Tickets can later be read by means of an optical character recognition (OCR) wand. Wanding provides feedback to the computer and collects payroll and production control data.

Since the bundle list is on file, only the coupon number and production order number need be read. The computer can identify all these for use as needed:

a. Product line.

b. Style or model family.

c. Model.

d. Style.

e. Color.

f. Size.

g. Quantity.

h. Time standard.

i. Bundle location.

Under an incentive system, the time which an operator earns from the coupons reported is converted to dollar earnings. This requires that the time which an operator spends working on an operation be recorded. Two methods are normally used:

a. Operators using a time clock record time of changing operations on their gummed sheets.

b. Supervisors give operators special coupons bearing the time they changed operations (usually accurate within 10 or 15 minutes). These coupons are OCR wand readable.

Some firms read the operator's coupon tickets on the gum sheet periodically during the work day. Payroll and work flow computations can be performed following each reading. This activity provides close monitoring of performance and flow. More current reports can be produced.

ROUTING SEQUENCES

Work placed in process normally travels a sequential route. Work passes through a succession of operations done at different work places. Engineers try to locate successive operations in adjacent or nearby work places as much as possible.

When all the styles running through a plant have the same operational sequence, the bundles of parts should follow the same route if the production line is in balance. Keeping the work moving through a stable sequence requires supervisory attention; therefore, supervisors are given more operators to work with. Sometimes styles placed in process together have varied operational sequences. This creates more than one route through the work places. Closer attention by both supervisors and bundle handlers is needed to make the styles follow different orderly routes without in-process delays. Supervisors usually are assigned fewer operators in a mixed style shop.

The same styles can be made by allowing bundles to travel alternate routes. At various points in the process, there may be operations which can be alternatively done without problems. In such cases, different routes may be followed owing to the fact that one operator happens to be ready for work instead of another.

Market trends indicate greater in-plant style diversity. More plants are experiencing complicated product mixes as time goes on. It is not realistic to expect the route of work through the process to remain essentially constant. Lines can be expected to change more in the future.

A flexible routing system requires:

a. More upfront production planning.

b. Loading for a balanced flow of work.

c. Implementing varied routings of style batches.

d. Making work-in-process reports more frequently.

e. Maintaining continuous flow.

f. Minimizing delay and misdirection of bundles.

WORK FLOW ROUTE DEFINITION

There are alternate routings and alternate assembly methods for bundles passing through operations which may be chosen for most apparel products. A system for describing work flow routes will be presented here. It is one which:

a. Easily adapts to alternative routes.

b. Precisely defines successive segments of activity.

c. Facilitates reporting of work in process.

d. Guides work loading.

The work flow routes require the assigning of subassembly route numbers and sequence level numbers prior to introducing the styles into a computer. The assigned numbers define operational sectors. These numbers usually appear together to define the activity which goes into making a product. They will be recorded routinely as part of the operation list for each style (see Exhibit 11.1 next page). Other systems of numbering are used. Leadtec will demonstrate a different monitoring system.

Components of products are frequently made in parallel, independent sequences of operations. These separate sequences ultimately merge into one route near the end of the process. Each of these sequences is identified as a subassembly route.

In any series of operations, it is usually possible to identify two relationships. The sequential relationship exists where certain operations must be done before others, and it would cause trouble to violate the sequence. The concurrent relationship exists where there are two or more operations that could be performed either simultaneously or alternatively started first.

The sequential relationship exists in a subassembly route between operational

EXHIBIT 11.1.

Style XXXX. Flow Chart of Operations
Establishes Rules of Order for Loading Work Upon Operations

ASSEMBLY ROUTES FOR COMPONENTS

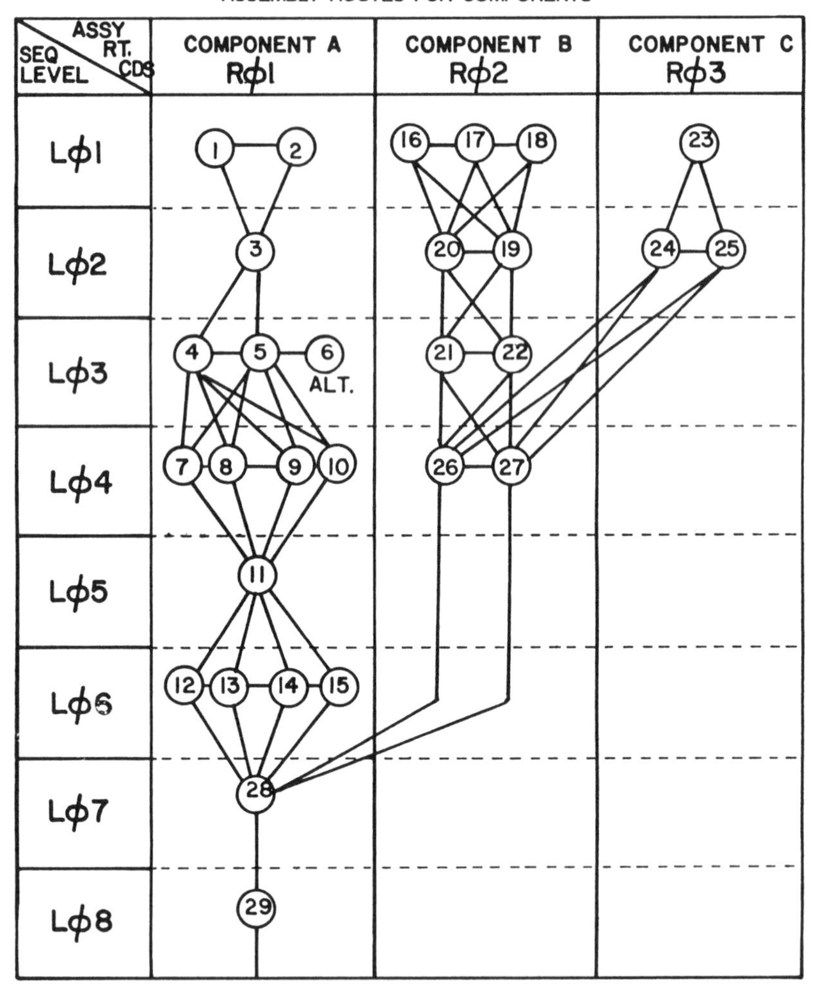

NOTES:

1. Assembly route flows must be coordinated to come together at points of confluence at the same time.
2. All operations in a segment defined by a sequence level and an assembly route must be performed on a bundle before any operation on the next level.
3. Operation 6 is represented as an alternate way to do the work of Operation 5.

sectors. The concurrent one exists on a sequence level within the sector.

Product operations must be examined to determine the parallel components they may construct. These operations are classified into separate series for each component. This defines the subassembly route. Each route is identified by number.

Operations which may be started alternatively or performed simultaneously will be grouped together. Then operations which must be done before others are arranged in numbered sequence. This places the operations on sequence levels.

The operations of each style should be flow-charted. Flow charting helps to reveal the component routes as well as the sequential and concurrent relationships. Computer people sometimes use this type of charting to determine the critical time path for a project.

Sequence levels are numbered from one to as large a number as needed. Every subassembly route begins on sequence level one and continues in order until the route merges with another.

Subassembly routes may be numbered as they are identified on a list. Care should be taken not to reidentify the same subassembly over and over again for different styles with a common component. The Leadtec system stores only the new operation descriptions when styles are entered. The system maintains a list of style operations which call out operation descriptions as needed to describe a style.

The subassembly route numbers do not have to be unique for every style. There are components which are very similar from style to style. There are similar collars on some sport shirts and windbreaker jackets. The operation sequences are much alike. Such collars can be defined as traveling the same routes.

The operation at the point of merging is always on the next sequence level after the one on which the last operation of the merging route resides.

There may be more than one sequence level within a work center, but there will be a whole number of sequence levels per center. All of the operations within the last level must be performed before bundles leave a work center. All of the operations in the next work center must be in subsequent levels.

Routes may run singly through work centers or may merge within them. Supervisory responsibilities for routes may be assigned as plant managers see fit. This route definition will be found compatible with established practices almost all the time.

A cardinal rule for each bundle is that all operations on one sequence level in one route must be completed before any operation on the next level is begun to prevent production bottlenecks.

Production is reported from defined operations. Inventory in process is reported the same way. This makes work-in-process reports unambiguous. When a bundle moves from one operation to another, it is counted as production. It has definitely progressed to a more advanced stage of completion. It definitely becomes inventory at the next operation (work ahead of the operator).

REPORTING FROM THE PLANT FLOOR

The piecework payroll and all working process reports depend upon exact data supplied from the plant floor. The information and data which are needed are essentially these items:

a. Operator doing the work.

b. Operations performed.

c. Production achieved.

d. Time spent doing the work.

e. Time spent in the plant all day.

f. Other activities identified.

g. Time spent on other activities.

The manner of reporting should be convenient for operators. The system should provide accurate data to the computer system. There are a number of ways to report activities, time spent, and production, if any. Each has some advantages and shortcomings.

Reporting from the plant floor may be done in real time mode or in batch mode. Techniques using a mixture of modes may also be possible.

The cost of each method of obtaining reports should be weighed against its effectiveness.

Reporting methods are covered to serve the purposes of an incentive payroll system, of monitoring work flow, and building a data base to support a plant loading system.

REPORTING TIME

Operators or supervisors report the time spent on significant events occurring during the workday. Usually the supervisor approves the event by initialing the operator's gum sheet. The significant occurrences for payroll purposes are:

a. Daily starting time.

b. Changing to a different operation.

c. Beginning an off-standard activity.

d. Ending an off-standard activity.

e. Beginning an indirect labor assignment.

f. Ending an indirect labor assignment.

Transferring to a different operation at a different rate of pay is simply changing to a different operation. Transfers at different hourly rates normally would be recorded as indicated in item f above.

The conventional coupon ticket and gum sheet used in apparel plants sometimes does not provide for clocking operation change times. It often includes room for only clocking of off-standard activities. One method in use is to have the operator carry the gum sheet to a job clock. There the time can be clocked in spaces provided along one edge. Recording operation change times is a similar procedure.

When the operation change time is not recorded, it is not possible to compute the operation efficiency accurately by operator. If this is not computed accurately, the average performance efficiency of the individual operator doing several operations can be determined, but the average performance efficiency realized by several operators doing the same operation cannot be correctly computed. The performance efficiency realized in producing the style cannot be computed correctly either. This means that extra costs generated by special styles cannot be identified for subsequent style costing activity. Operation change times are therefore important events in a time recording system.

Accepting that the times of all the significant events affecting operators should be recorded, it becomes necessary to install a convenient way to log these events. Various clerical and mechanical approaches have been and are now used. Today several more sophisticated electronic approaches are available to improve activity time accuracy.

A few methods follow in the list below:

a. Have employee timekeeper keep a log for all operators.

b. Have operators write down their own event times.

c. Have operators go to a time clock to record times on a card or gum sheet.

d. Have operators clock in and out and have supervisors clock events in between these times for all of their operators.

e. Have supervisors provide preprinted time-bearing coupons to the operators to be placed on gum sheets separating sequential groups of operation work coupons.

f. Have operators record events using keypads at work places wired to computer input devices.

g. Have operators record the events using wireless transmitting devices.

h. Have coupons printed with bar codes and optically recognizable characters and read gum sheets using OCR scanners.

A central problem in all time recording methods and systems is how to verify the record. Supervisors are usually held responsible for time verification. Some operators have been known to cheat. They have friends clock them in on time mornings when they are late and try to record as much off-standard time as possible for themselves. When an operator's true working time is artificially shortened by declaring some of it to be off-standard, their time appears to be more efficiently used than really it is. Efficiency directly affects the income of operators on incentive. Inattentive supervisors can contribute to inaccurate recording habits and to deterioration of the incentive system.

The most difficult part of the above listed systems is associated with time verification. This is the reason supervisors are usually involved in the time recording. Timekeepers involve a continuing indirect labor cost. Most apparel firms do not use timekeepers. A few apparel firms use an honor system. Operators declare their own time. This has led to abuse and distortion of time records.

Causing operators or supervisors to walk to time clocks to clock operation changes and off-standard activities loses time and causes annoyance. It does not ensure accurate recording unless perhaps the supervisor sees and verifies by signature that the time of the event is recorded near the true moment of occurrence.

The participation of the supervisor requires his or her presence at the operator's workplace when the operation changes. The supervisor could use time more effectively if changes could be preplanned. Then changes could be carried out while the supervisor attends to some other matter. The need for verification interferes with this practice.

Often supervisors do not let it interfere; they let the operators do the recording of time unobserved, and then they verify it later accepting the operator's word that the time is correct. This is very loose verification but it probably has some value. Operators probably would find it easier to lie to an uninspected record than to a person with authority, if they felt equivocal about lying to begin with.

The technologically sophisticated methods of recording event times as computer inputs do not offer any improvement in the area of verification. The advantages of these methods are mainly speed and accuracy of input.

Real time reporting affords a means of electronically recording the event times at the workplace at the moment of occurrence. In most systems card readers or scanners are located at workplaces and connected by wires to the computer. Operators can key or read in their operation numbers, bundle numbers, and starting times. This serves the purposes of both payroll and production control systems. It provides an immediate and accurate record. However, it is only loosely verifiable. Implausible reports can be investigated.

Batch time reporting involves accumulation of data for a period of time and then entering the data at one time into a computer. This can be done using a terminal or by scanning the data with special optical and electronic devices. Printed input media would be presented to these devices in optically recognizable characters or bar codes. Coupon tickets could be printed for OCR or VR scanning. These coupons can be accumulated on gum sheets and can be read into a computer file periodically using an OCR wand. Loose verification of the times could be provided by the supervisor, as with the other methods.

Real time reporting has the advantage of being immediate. All files and reports are always up to date. The data source units which are used with the REACT, Eton, Gerber and other systems were expensive when first developed. Technological advances have reduced the installed cost. This makes the use of on-line systems more feasible.

Batch reporting using OCR wand input is less expensive to install. The computer operator wand reads and keys in the data. This is a small continuing daily expense. A discussion of batch input methods using OCR wands follows.

Using an OCR wand to input data is a convenient way to record operator, time, and production. Both time and production inputs are read together at the same time. The system used in most apparel plants to report production is to have operators clip pre-prepared coupons from bundle tickets and then stick the coupons on their individual operator gum sheets.

The operator is identified by an individual coupon showing a payroll number. A periodic supply of these are computer produced and given to each operator. This special coupon is the first one placed on the gum sheet each day.

One system uses special preprinted coupons to report time. The computer produces a supply every two or three days. These special time coupons are stocked in the supervisor's locked cabinet. Coupons are printed for the start of every five-minute interval throughout the workday. Including an hour of overtime, 108 different OCR coupons are used.

At the start and end of every workday, supervisors give out time coupons to the employees. Anyone late in the morning has to ask for a coupon to represent starting time.

Supervisors authorize working through lunch. The supervisor could give an operator a special coupon or note the event on the time sheet. Otherwise, the standard lunch time period would be automatically deducted by the computer from the total time of the workday.

Each time an operator changes an operation, he or she either clock in or ask for a time coupon.

Coupons are grouped together on the gum sheets by operation. These groups are separated by a time coupon or a time clock printout which indicates the time that operations were changed. The time spent would be associated with the exact amount of work and the operation. A similar procedure is used for machine downtime, wait on work, etc.

Even if there is a preprinted coupon-fed system for keeping track of time, it is recommended that each employee use a conventional time clock for clocking in and out daily. This procedure complies with federal wage and hour laws.

REPORTING PRODUCTION

When a production coupon is read on a gum sheet, the computer needs the production order number, the operation number and the coupon number in order to reference other relevant information. The style, quantity and time standard are on file.

The bundle list is in file. This list gives the size, color, and bundle quantity together with the bundle serial number. It is cross-referenced to the style number and the production order number. The sum of the bundle quantities on the bundle list is the total quantity of the production order (or cutting ticket).

The style number may be used to refer to the operation list which is on file in the computer. This provides all the information concerning the operation ready. Operation lists the rate file. The rate file records the method and is the source of the correct time standard.

The product, the operation, the quantity produced, and the time interval spent doing the work are all fully identified by the system when each coupon is read.

MONITORING WORK FLOW

Production planning and scheduling activity are intended to provide a balanced flow of work through process. This means that work is planned for each operator to

ENGINEERING, TRACKING WORK IN PROCESS, MEASURING PERFORMANCE

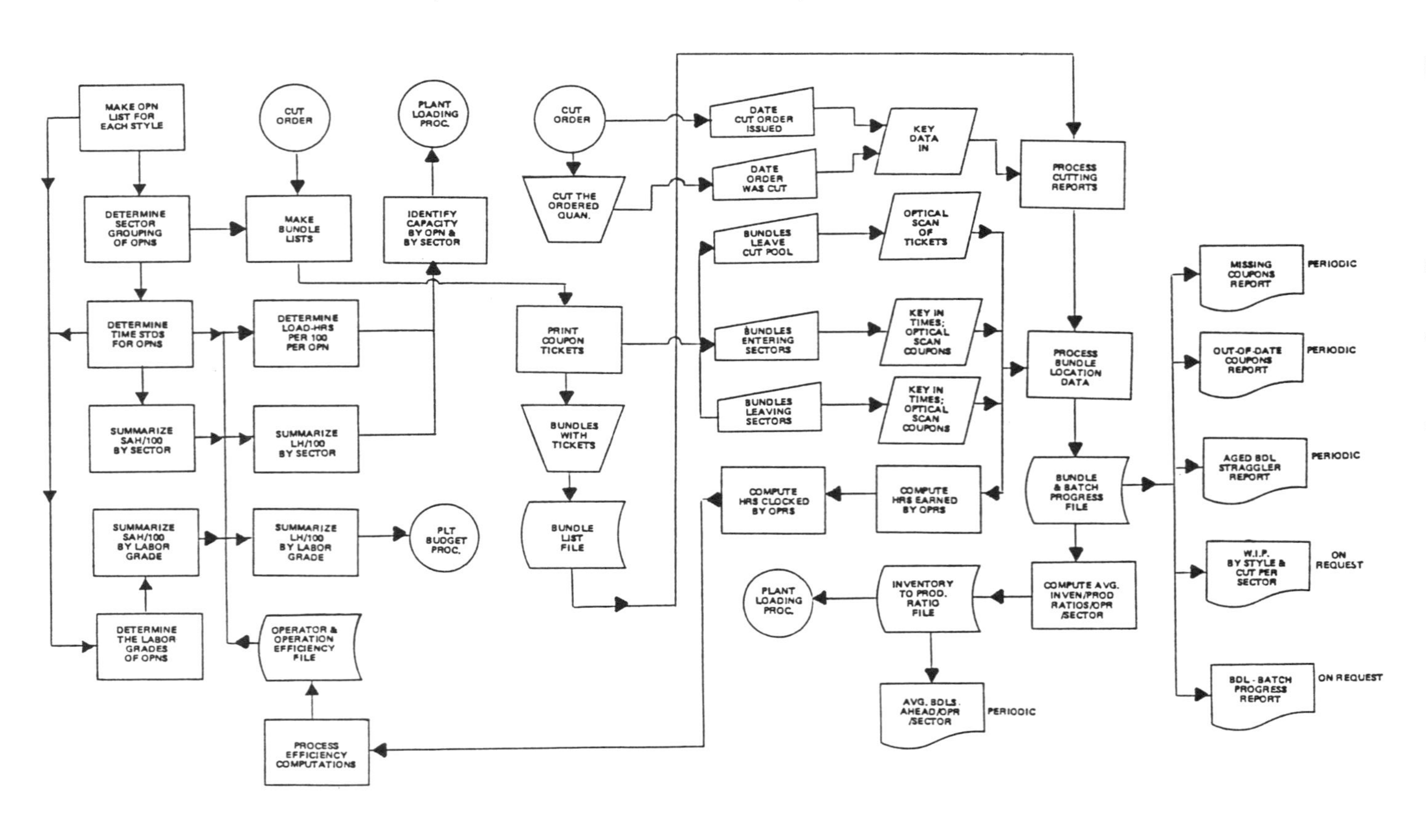

balance inventory within the production line. The fact that it is planned does not absolutely insure that the work will flow evenly through all operations. Supervisors are responsible for maintaining the work flow.

A production control report monitors the work flow in order to detect where inventory increases and decreases in process. Plant managers and supervisors evaluate the effects of changes. Regular reports are required to show change. They should be easy to interpret. Some programs show upper and lower inventory limits for key operations. These limits are used to control production.

Control limits are also used to maintain a reasonable amount of inventory ahead of every operation as well as to avoid inventory buildups. An inventory buffer keeps all operators occupied and tends to absorb production delays. Temporary stoppages do not travel from operation to operation in a ripple when there is enough inventory in process to dampen them.

Too much inventory may exceed storage capacity, increase inventory holding costs, and increase the chances that operators at following operations will have to wait for work. When this happens, labor costs increase.

Appropriate reports, their interpretation and proper action to control the flow of production will be described in the following subsections.

PRODUCTION AND WORK-IN PROCESS REPORTS

Normally, five reports may be triggered from the computer periodically. These reports are based upon the data reported from the plant floor. They are named:

Bundle and Batch Progress Report

Report may be prepared daily for each supervisor. The report is available as a document by inquiry at a terminal.

It lists the following:

a. Subassembly route.

b. Sequence level.

c. Production order number.

d. Bundle number.

e. Bundle quantity.

f. Operations not yet done on each bundle on each sequence level.

g. Sums of bundle quantities by subassembly route and sequence level.

h. Sums of bundle quantities by work center.

i. Production since last report.

j. Week to date production.

All the bundles waiting in each operation or sector are grouped by production order number. The report is organized to show where the various parts of a given production order are located. The report also shows how bundles move from one stage of completion to another. It is designed to locate bundles within each sewing section.

There may be more than one production order per production sequence level. Production orders normally contain more than one bundle. Bundle quantities at an operation constitute the inventory there. The total bundle quantities in all the sectors within one work center comprise the total inventory in process in the work center.

Production and Work-in-Process Summary

The summary gives the total production from each line and the total inventory in process located within the line. It gives an inventory summary for each work center. The report does not identify bundles or production orders but deals in total units. Normally this report is for each line within the plant. Plant inventory totals are also calculated.

The heading of this report identifies the plant and the time and date of the report. The report lists for each assembly route and sequence level the following:

a. The production from each sector.

b. The work-in-process in each sector.

c. The total work-in-process for the work center.

d. The inventory in all sectors are totaled for the whole plant.

Straggler Report

Report lists those bundles which have remained too long in the same place (i.e., 2 days). It is usually an exception report. When any bundle has been reported on two bundle and batch progress reports in the same area, it can be classified as a straggler. The body of the report lists for each assembly route and sequence level number the production order number, the bundle number, the bundle quantity, and the operations not done.

Operations Out-of-Sequence Exception Report

Report identifies bundles for which operations have been reported done out of sequence. This would be either a reporting mistake or an operating mistake. The report identifies out of sequence bundles by production order, style, bundle number, quantity, and work center. Bundles which have an operation reported finished which skips a required earlier operation are reported. This report may indicate coupon ticket holding.

Inventory-To-Production Ratio Report

Report may be prepared daily. The report may be produced as a document or it may be available for inquiry at a terminal. Supervisors and plant managers use this report to evaluate how well the work flow is balanced.

This is prepared for the plant as a whole and is analyzed by work center and sector. The heading of the report identifies the plant and gives the time and date. A report lists the following:

a. Work Center.

b. Sequence Level.

c. Assembly Route.

d. Inventory in each sector.

e. Production from each sector.

f. Number of operator-days in each sector.

g. Current average ratio per operation in the sector.

h. 20-interval moving average ratio per operation in the sector.

If the average ratio per operation varies widely from sector to sector, this indicates that the work flow has become unbalanced. This is evidence that the inventory has either accumulated or diminished in different sectors, and may require closer supervisory attention.

A moving average ratio for, perhaps, a four week period may be kept to evaluate whether the current ratio is varying from the moving average.

INTERPRETING WORKFLOW REPORTS

The different reports previously outlined reveal different aspects of the problem of maintaining the workflow. This section deals with who should receive each report and how it should be interpreted, if such reports were actually in use.

The receivers are indicated for the following reports:

a. Bundle and Batch Progress Report

   (1) Supervisors.

b. Production and Work-in-Process Summary

   (1) Supervisors.

   (2) General foreman.

   (3) Plant manager.

c. The Straggler Report

   (1) Supervisors.

   (2) General foreman.

   (3) Plant manager.

d. Operations Out-of-Sequence Exception Report

   (1) Supervisors.

e. Inventory to Production Ratio Report

   (1) General foreman or superintendant.

   (2) Plant manager.

   (3) Production planner.

EXAMPLES OF REPORT FORMATS

BUNDLE AND BATCH PROGRESS REPORT

Work Center ______ Supv. ______ Time ______ Date ______

Assy. Route	Seq. Level	Prod. Order	BDL No.	BDL Quan.	Operations Not Done
Work Center Inventory					XXXXXXXXXXXX
Work Center Production					XXXXXXXXXXXX

PRODUCTION AND WORK-IN-PROCESS SUMMARY

Mfg. Plant ____________ Time ____________ Date ________

Work Center	Inven. In Proc.	Wk Ctr Prod	Work Center	Inven. In Proc.	Wk Ctr Prod

EXAMPLES OF REPORT FORMATS

STRAGGLER REPORT

Work Center ________ Supv. ________ Time________ Date _______

Assy. Route	Seq. Level	Prod. Order	BDL No.	BDL Quan.	Operations Not Done

INVENTORY TO PRODUCTION RATIO

Mfg. Plant _______________ Time ____________ Date _____________

Work Center	Average Inventory	Daily Prod.	Raw Ratio	Oprs. Wkg.	Curr. Avg. Ratio/OPR	Moving Avg. Ratio/OPR

Bundle and Batch Progress Report

Report is designed to enable supervisors to locate and to follow the progress of bundles and production order batches through their areas. They will be able to discover:

a. Which bundles are in each part of their production line by production order.

b. Which operations the bundles must pass in order to move to the next sequence level.

c. Rate of production being achieved for each production batch from each sector, and from the work center.

d. How the work flow corresponds to schedule.

e. Locations of inventory accumulations or depletions owing to imbalances to the work flow.

Production and Work in Process Summary

The summary gives an overview of the results of the plant and production lines daily. The report is a moving update of the latest full day's results. It describes the current daily rate of production from every work center and from the plant. It indicates whether or not inventory in process is building up or diminishing anywhere. It affords a comparison of actual daily production rates to planned daily production rates, as well as a comparison to the daily production rates seen two hours earlier.

Straggler Report

Report may be produced daily. It is also a moving update report. It tells what bundles have remained on one sequence level for eight hours. It enables supervisors to find and expedite any bundle that has been overlooked or accidentally bypassed. Because this is an exception report, it reports nothing that is unimportant. It there are no stragglers, there is no report.

If any batch in process is deliberately bypassed for several days in order to move other work ahead, a Straggler Report keeps track of the location of every bundle. It is a useful tool for getting the bypassed work back in the main stream of production.

Operations Out-of-Sequence Exception Report

Report identifies the production order, style, and bundle number for which an operation has been performed out of sequence. The bypassed operation is identified. The report is sent to the supervisor in the area in which this occurred. The supervisor would then find the bundle and decide what action to take.

Operations performed out of sequence are exceptions reported to the supervisor. The report occurs when the coupons have been lost, when an operation has been bypassed, or when coupons are misappropriated. All these conditions require supervisory attention.

Inventory to Production Ratio Report

Report evaluates the state of work in process so as to estimate what inventory might be expected at the end of the production line. The ratios evaluate in process work turnover relative to production. Ratios serve as factors to compute the time for a bundle or batch to pass through the process.

The inventory to production ratio makes a comparison between the inventory maintained ahead of an in-process operation and the production from the operation. The report computes an average for each production operation.

Correct in-process inventory varies with circumstances. Enough work is needed ahead of operators to absorb delays. Too much inventory will clutter the plant and increase costs. There is a quantity of inventory which presents no problem to the supervision. Some latitude for practical variation is needed.

If the time standard for the operation is 1.0 hour per 100 and the operator performs at 100% of standard, it is easy to compute the length of time the operator will be occupied by the in-process inventory. As an example, let the standard bundle contain 25 units. One bundle supplies 15 minutes of work. Two bundles amount to a half hour of work for one operator. Then four bundles are required to provide a half hour of work for two operators doing this same operation.

The time it takes a bundle to pass one operation in a sector on average for a time period during which production and inventory were counted would be calculated as:

$$\text{Hours transit} = \frac{\text{Inven.}}{\text{Prod.}} \times \frac{\text{Bdl.}}{\begin{array}{c}100\\ \underline{\text{Bdl.}}\end{array}} \times \frac{\text{SAH/100}}{\text{opn. perf. eff.}}$$

$$\text{Hours transit} = \text{Ratio} \times 100 \times \text{LH/100}$$

The sum of the times passing each operation in a sector is the whole time it takes a bundle to travel through the sector. The number of operators assigned to the sector is not significant in this calculation.

Some complications to this simple system may arise within the assembly sector (i.e., parts). The operations may be performed in random order or simultaneously. It is difficult to calculate the inventory ahead of each operation. Under this condition, averages are used for each sector. The average ratio is determined for the sector, and when needed, the average bundle size may be computed.

CONCLUSION

There are multiple reports available from any production/payroll/in-process data base. The foregoing examples have been drawn from several commercially available software systems. The basic data within the manufacturing computer systems cover activity from the receipt of raw materials through the movement of finished goods to the distribution center.

Piece goods and trim aging controls, interaction between manufacturing, financial general ledgers and marketing are built into most systems packages today. In the next chapter, some further aspects of the payroll systems will be examined.

12

Payroll

INTRODUCTION

Most apparel firms use a payroll system to pay direct labor employees. The system is designed to motivate people to try harder because they are paid for additional effort.

As outlined in the preceding chapter, piecework payroll systems require the collection and processing of data which can be of great potential use in manufacturing. The development and use of an integrated, computerized manufacturing system in conjunction with the payroll system expands the range of application of the data enormously.

A piecework payroll system is a fundamental data source. It is a mine of information which can be applied in such areas as product costing, productivity measurement, performance evaluation, production planning, scheduling workflow control and more. The power of the computer to access data, to process large amounts of it and to perform lengthy calculations quickly extends the useful range of the data base afforded by the payroll system a great deal. This portion of the text covers payroll and some byproduct engineering and costing reports.

PAYROLL DATA INPUT

Most apparel firms use bundle tickets with work ticket coupons to record a completed job. These coupons are usually affixed to a "glue sheet."

Data from the plant floor may be entered by scanning coupon tickets with an optical wand, or by key punching the data from the "glue sheet." These production coupons identify the product and the time allowed for the job. The operator's worksheet identifies the operator and starting and finishing times. As described in the preceding chapter, coupons are stuck on the worksheet in the real sequence of events as follows:

a. Operator identification coupon first on the sheet.

b. Starting time coupon.

c. Production coupons for the first operation performed.

d. Operation change time coupon.

e. Production coupons for the second operation performed.

f. Finishing time coupons.

Payroll coupons can be read into the computer system in various time spans (i.e., 2, 4, or 8 hours) in order to monitor the workflow. This input serves the payroll as well, although this frequency of reading is not necessary to compute the payroll. One time daily is normally sufficient for payroll.

Implementing payroll systems requires that other files of data be built. Typical files are:

a. File cross-referencing style numbers and operation list numbers.

b. Operation list for the style entering process.

c. 4M methods and standards file (or equivalent)

   (1) Unique composite operational standards.

   (2) Standards for methods composed of standard data elements.

d. Bundle list of all bundles made from the cut of the production order with the coupon ticket serial numbers.

e. The base hour rates of pay in a file established for labor grades by operator. This file relates the labor grade to the operator's name.

PAYROLL COMPUTATIONS

Payroll computation's fundamental objective is to write paychecks, to compile and to report labor costs. Once the data for payroll is collected and entered into the system, the analysis and the computation can begin.

Payroll processing is divided into two sections:

a. Source-to-gross computations.

b. Gross-to-net computations.

Source-to-gross computations begin with the detailed reports from the plant floor. Direct, indirect and overtime hours are distinguished. Piecework earnings in hours are determined. Total timework hours are calculated. Time is converted into gross dollars earned.

Gross-to-net computations begin with receipt of gross dollars earned. Labor costs are broken down into direct, indirect, overtime, and makeup. Withholding tax and other deductions are made; pension fund and other reports are produced.

Further explanation of the computions involved are presented in the following subsections.

SOURCE-TO-GROSS PAY

The source-to-gross process takes the basic reports of time spent, activities, and production and converts them into gross earnings.

The direct labor is paid according to a piecework system. Reports from workers in this category are normally kept separate from reports on workers classed as indirect labor. Different procedures are used to calculate the two payrolls.

The hourly rates of pay for direct nd indirect workers may have been derived by job evaluation or negotiation. Regardless of how these rates are analyzed or organized, the computer system should be able to deal with hourly rates as though they were individually assigned to operators.

Indirect labor time is clocked on individual time cards. The time is summarized and multiplied by an individual employee's hourly rate. This hourly rate may fall between the limits of a labor grade but would be individual for the employee.

Base hourly rates for direct labor can be established objectively by skills and training time involved. Such rates can be established for operators who learn only one job skill. After time passes, operators may learn more than one operation.

An operator who learns several operations with different base rates associated with different skills will require a record showing the additional abilities. Pay adjustments must be made for operators reassigned from a job with a high base rate to a job with a lower one. It may be appropriate to pay the operator the highest base rate for all the operations in his/her skill repertory as soon as the operator had been able to achieve 100% of the standard for the highest base rate operation. Management should establish a policy to cover this circumstance at a top level; otherwise, the problem will be dealt with differently from plant to plant within the firm.

Time standards used in computing direct labor and piecework earnings are recorded in a 4M Methods and Standards (or equivalent) file. The operation list for each style references the appropriate time standard for each operation. The coupon on the bundle ticket references the pertinent operation on a correct style operation list. Time standards are always made accessible for the payroll computation by the coupon form the bundle ticket. It is an optional matter whether the time standard is actually printed on the coupon or not. Some operators like to see the number of units they need to produce each hour, others prefer to see the dollars they are earning as they work.

The quantity, which has been put into each serially numbered bundle, is recorded in a bundle listing file for each production order. The coupon on the bundle ticket references this bundle list for the production order and finds the correct quantity for each bundle.

Each coupon is used to multiply the quantity by the time standard to obtain the hours earned. Coupons from all bundles passing the same operation in a continuous time period are summed to determine the total hours earned during the period. The actual continuous time spent is then determined by taking the difference between the starting and stopping times less off-work time such as machine delay. The hours earned are divided by the actual hours spent to obtain the performance efficiency for the operation.

These items of information should be recorded for each group of coupons:

a. The operator identification.

b. Operation number.

c. Style number.

d. Production order number.

e. Hours earned on the operation.

f. Hours spent on the operation.

g. Performance efficiency achieved.

The hours earned and actually spent on different operations throughout the work day and work week, are summarized by operation for the whole period. The quotient of these sums is the operator's overall weighted average performance efficiency for the period. This data may be compiled by operator for a predetermined time period (i.e., a quarter) to record skill progress.

The operator's <u>actual</u> hourly rate of earning will be the <u>base</u> hourly rate multiplied by the overall <u>performance</u> efficiency. The operator's dollar earning from piecework will be the <u>hours spent</u> multiplied by the <u>actual</u> hourly rate. This gives precisely the same dollar earning as multiplying the <u>hours earned</u> by the <u>base</u> hourly rate.

An example will show how the operator's performance for a period of time is an average weighted by the time spent. Corresponding earnings are computed using a $4 base rate.

TABLE 12.1
ESTABLISHING PERFORMANCE EFFICIENCY

Operation	Hours Earned	Hours Spent	Perfor. Effic.	Actual Hour Rate	Dollars Earned
A	15	12	1.250	$ 5.00	$ 60.00
B	6	5	1.200	4.80	24.00
C	9	10	0.900	3.60	36.00
D	9	7	1.286	5.14	36.01
E	8	6	1.333	5.33	31.99
Sums:	47	40			$188.00
Avgs.:			1.175	$ 4.70	

The 47 hours earned divided by the 40 hours spent equals 1.175 which is the weighted average performance efficiency. The actual average hour rate is $4.70 which is an average weighted by the hours spent.

Operation performance efficiencies vary with respect to the operator average efficiencies. They are not difficult to analyze since the data is already separated in the first place. It will be shown later how operator and operation performance efficiencies may be separately compiled and used.

Off-standard time would be declared on the operator's work sheet as each instance occurs. The identity and the amount of off-standard time must be verified by the supervisor. Categories of off-standard time must be pre-established. The supervisor should specify the hourly rate of pay for timework. This must have to conform to an established company policy.

Total hours worked may be partly piecework and partly timework. Payment to an operator for off-standard activities is the product of the verified time and the specified hourly rate. Different occasions might entail different hourly rates, such as the minimum hourly rate or the operator's average hourly rate. Off-standard pay for any time period can be the sum of the products corresponding to the individual occasions.

Payroll make-up is an excess direct labor cost. It is an operator's loss of earnings as much as the hourly rate of pay which has to be paid according to law or agreement. It is called "make-up" by the company and paid even though it is not really earned.

When computing make-up, you must first determine which hourly rate applies to the case. Two kinds of make-up are defined with reference to the hourly rate of pay:

a. Make-up to minimum.

b. Make-up to average.

Make-up is the difference between operator timework and piecework earnings computed for the same period of time. The timework earning is the number of hours spent multiplied by the hourly rate guaranteed for that time. The piecework earning is computed, as explained before, as the summation of the products of the clipped coupon quantities and time standards which is multiplied by the base hourly rate established for piecework. Make-up exists when there is a positive difference obtained by subtracting the piecework earnings from the timework earnings of an operator.

There are different categories in which make-up may be placed. These are:

a. Make-up incurred by experienced operators on regular operations.

b. Make-up incurred by experienced operators learning new operations.

c. Make-up incurred by trainees.

After categorization as above, make-up may be analyzed by work center within the plant. A training center, when it exists, is regarded as a separate work center.

Computing payroll make-up is part of the source-to-gross payroll processing. The compilation of make-up summary reports may be included in the gross-to-net processing, but it would use data furnished from the source-to-gross processing.

Payroll make-up is accounted for in dollars as an excess expense. It is not dealt with as an expectable loss of direct labor capacity (like off-standard time or absenteeism). Make-up is not used in production planning. There are two reasons for this practice:

a. Make-up is escapable, correctable and is reported so that it may be eliminated.

b. Make-up corresponds to inefficiency, for which there already is a measure--the performance ratio.

Overtime hours which may be spent are multiplied by the established overtime premium. The dollars are added on to piecework and timework earnings.

The final results of these operations are to compute the gross dollar earnings of operators.

GROSS-TO-NET PAY

Gross-to-net payroll calculations begin with the input of gross dollar earnings. It updates individual payroll records and ends in writing paychecks.

These earnings are broken down into indirect labor, direct labor, make-up incurred by piece workers and overtime in both direct and indirect categories.

Total piecework hours in dollars paid are available for each pay period. These data are used to calculate the thirteen week moving average hourly earnings for each operator. All tax and other deductions from gross earnings are made and accounted for. A statement of deductions is printed on the stud with the paycheck.

The following items of information are the normally maintained payroll records for every pay period on every employee in every plant:

a. Plant.

b. Employee.

c. Clock Hours.

d. Overtime Hours.

e. Standard Allowed Hours.

f. Timework Hours.

g. Off-standard Hours.

h. On-standard Hours.

i. Allow Hours.

j. Earned Pay.

k. Timework to Minimum.

l. Make-up to Minimum.

m. Separate Make-up.

n. Vacation Dollars.

o. Sick Pay.

p. Holiday Pay.

q. Gross Pay.

r. Net Pay.

s. Bonus Pay.

t. Overtime Premium.

The gross-to-net pay system provides for deductions identified as follows:

a. Credit Union.

b. Pension Fund.

c. Special Political Contributions.

d. Loans.

e. Monthly Investment Plan.

f. Purchases.

g. Savings.

h. Hospital Insurance.

i. Life Insurance.

j. Family Life Insurance.

k. Union Dues.

l. Bonds.

m. Charity.

n. United Fund.

o. Christmas Club.

p. Garnishment.

q. Negative Deduction Corrections.

r. Miscellaneous.

s. Federal Withholding Tax.

t. FICA Tax.

u. State Withholding Tax.

Maintenance of the master employee files is carried forward collaterally with the operation of the gross-to-net payroll processing. This activity provides for:

a. Setting up new employee files with

   (1) Personal data.

   (2) Payroll data.

b. Updating files current for

   (1) Payroll release.

   (2) Quarterly report processing.

   (3) Annual report processing.

c. Computes and records payroll deductions for

   (1) Taxes.

   (2) Authorized purposes.

   (3) Garnishments.

d. Balancing and crossfooting payroll totals.

The gross-to-net payroll processing comprises computing the pay, making the deductions, and preparing a payroll release report. These data are transmitted to the plant location where checks are printed out. Also, the payroll release report is printed furnishing the plant management with a listing of employees paid and checks issued.

In conjunction with the check writing, W-4 data is updated for annual reports. The quarterly report of taxable earnings, Report 941A, is prepared for the federal government.

The gross-to-net processing also incorporates the procedures to calculate the 13-week average hourly earning of each employee, and to calculate each employee's over-all 13-week efficiency. Reports of these results can be produced on listings organized by plant, department, work center, and employee name. They can be printed remotely at plant locations or at corporate headquarters.

The system also can produce further listings upon request:

a. Complete employee listing.

b. Birthday listing.

c. Deductions for each employee.

d. Employees eligible for service awards.

e. Employees eligible for holiday pay.

f. Employees vacation gross pay.

g. Employees on pension plan.

Any payroll system must provide for the various maintenance, updating, and data movement operations which are useful in data processing.

DATA BASE FILES CREATED

The files described here are those which are considered to be needed for various plant applications. Some of this information is already made available by any existing base payroll system. Probably some modification to any existing payroll files is necessary to provide all that is suggested here. The payroll data thrust of this portion is to describe a data base which will serve the interests of plant managers and those planning scheduling and controlling production, using an integrated system to carry out most of the detailed procedures.

These files are maintained following calculation of the payroll. Updating them is a by-product of the payroll system.

PAYROLL JOURNAL

A payroll journal is the daily record of the transactions that took place in the source-to-gross payroll computations. The journal records data so that plant supervision can explain and justify payroll results to operators as soon as question may arise.

This journal can help avoid further misunderstandings also. For example, production coupons of the sort that are read by OCR scan of bar codes or numerical codes may not have as full a description printed on them as noncomputerized coupons do.

Operators may become uneasy with not being able to read all that they could before. The operators will want to check upon how their earnings are being recorded, and they should be able to do so. The Payroll Journal should be designed to provide proper explanation.

This journal should be accessible daily upon inquiry if it is not actually printed out. It lists operators by name and in alphabetical order, groups by work center and department within the plant. The date of the day and the week ending date appear in the heading together with the identification of the plant. For each operator there is an analysis of piecework results by operation and the summation of piecework and excesses for the day.

The piecework would be analyzed as follows under each operator name:

a. Operation number (one line for each operation number).

b. The time standard.

c. Quantity produced.

d. Hours earned on piecework operation.

e. Hours spent on piecework operation.

f. Performance efficiency on the operation.

g. Actual hourly rate of earning in dollars.

h. Piecework earning from the operation.

Daily summaries would be provided for the following items under each operator's name:

a. Off-standard time.

(1) Paid at minimum.

(2) Paid at average.

(3) Paid at certain percentage of average.

b. Off-standard dollar earning.

(1) At minimum.

(2) At average.

(3) At percent of average.

c. Overtime hours worked.

d. Overtime premium earning - total dollars.

e. Other excesses paid - total dollars.

f. Total hours spent (clocked).

g. Total hours earned on piecework.

h. Total gross dollar earning.

This journal is also the basis for a file. It is the source of data which enables the performance files to be produced for reference. The analysis of performance is treated in a later section.

INDIVIDUAL OPERATOR RECORDS

Two files contain complete employee information and work history:

a. Employee personnel records.

b. Individual employee payroll records.

These files should cross-reference one another so that both could be accessible at once if desired. They would be kept in the location where the gross-to-net payroll program operates. It should be possible to download any of the data to plant locations when needed.

Individual employee payroll records would preserve the daily summary values from the Payroll Journal. These do not include sufficient detail to trace the computation of pay, but would present the main results.

Individual employee personnel record files include:

a. Employee number.

b. Employee name

c. Plant.

d. Department

e. Work center.

f. Address.

g. Telephone number

h. Social Security number.

i. Birth date.

j. Sex.

k. Marital status.

l. Tax exemptions, federal.

m. Job code.

n. Pension code.

o. Date employed.

p. Listing of operations able to perform.

q. Average hourly rate (13-week moving average) .

r. Base hourly rate.

s. Minimum hourly rate.

t. Reference to commendations.

u. Reference to censures.

The details of censures and commendations may be kept in another file. Individual payroll record include these items:

a. Employee number.

b. Employee name.

c. Total hours spent (clocked).

d. Total hours earned.

e. Weighted average operator performance efficiency (current).

f. Actual piecework earnings.

g. Off-standard hours.

h. Off-standard pay.

i. Overtime premium (posted daily when daily guarantee exists).

j. Other excesses.

k. Gross dollars earned.

PERFORMANCE EFFICIENCY

Payroll computations make possible computerized analysis of the real performance efficiency in three distinct and useful ways. It is possible to obtain on both the current and moving average basis the following performance ratios:

a. Operator weighted average performance efficiency.

b. Operation weighted average performance efficiency.

c. Style performance efficiency.

These ratios represent the efficiency with respect to standard. They are the averages experienced weighted by the time spent gaining that experience. They are the best and most useful evaluations of performance available.

It is always possible to think that things might have been done better in the past or might be done better in the future. This is frequently just subjective impressionism. The level of achievement experienced is the best point of reference to which future plans for similar undertakings may be compared. Unless applicable resources can be identified which are superior to those used in the past, it cannot be reasonably expected that a new plan will do better than before. Following this line of reasoning, the pertinent performance efficiency experienced has to be considered along with the way it was achieved before.

The composition and application of performance efficiency files are discussed in the following sub-sections.

OPERATOR PERFORMANCE EFFICIENCY

For every working day there is a current operator performance efficiency value computed. It is an average of the performances achieved on each operation weighted by the time spent working on each one.

The operator performance efficiency indicates how well the operator works and earns. It furnishes supervisors with a means of evaluating the performance of their operators.

Operator performances grouped by work center in each plant should be listed and furnished to the supervisors daily. The information can be used as an effective management tool to motivate people and perhaps stimulate constructive competition.

These ratios may be computed on a 13-week moving average basis as well as for the current pay period. The average performance ratio would correspond to the moving average hourly earning.

There are some managers who feel that a four-week moving average on operator performance is adequate. This value is justified as being more related to current performance. If the 4-week moving average has merit, then it is recommended that both 4-week and 13-week averages be kept. Then current performance can be compared to quarterly-term performance.

Moving average may be dealt with two ways. If the current weighted averages entering the moving average are daily averages, a 65-day moving average should be computed. Any holidays or vacations would be current once each week. The current value could be computed weekly instead of daily and enter a 13-week moving average. Variations within the week would not be revealed in this case.

When selecting operators to do operations, it is desirable to know the operator's performance on prior operations and to know the amount of experience the operator has acquired doing them. It is important to preserve the records of the different operations, the performance efficiencies, and the accumulated hours of experience on each one. These data provide the rational basis for assigning operators to new operations.

The operator efficiency file serves the needs of both:

a. Supervisory control.

b. Operator assignment during work loading.

OPERATION PERFORMANCE EFFICIENCY

Operation performance efficiency rates plant experience for a particular operation. More than one operator may have worked on a given operation. The current weighted average performance efficiency is computed averaging the times spent by these different people.

Each operator's earned hours is computed by multiplying the quantity produced by the time standard. Each operator's clocked hours for working on that same quantity is recorded on the incentive payroll report, the gum sheet for coupons.

The earned hours for an individual operation are summed for each payroll period. The hours clocked on the individual operation would be separately summed for each payroll period. The quotient of the total hours earned is divided by the total hours clocked to compute operation efficiency for the period.

This efficiency takes in the performances of all the experienced operators working on the operation. It is advisable not to include the hours spent by trainees and utility operators.

Computing moving average performance efficiencies for operations provides stable, reliable operation efficiencies, useful in evaluating past performance and estimating future obtainable performance. A 13-week moving average operation efficiency would cover the same length of time for this ratio as would be used for computing average hourly earnings. It would contain time contributed by every operator able to perform the operation.

The sums of hours earned and hours clocked would have to be retained for 14 weeks in order to maintain this moving average for each operation. The current week's sums would be added into the total of them all. Sums of the 14th week back in the past would be subtracted out of this grand total. Then the 13-week sum of hours earned would be divided by the corresponding sum of hours clocked to give the moving average operation efficiency.

Operation efficiencies should be determined for each operation of each style passing through process. These efficiencies should be compiled to correspond to the list of operations for each style. In this manner, the experienced performance efficiencies are readily and conveniently accessible for use in loading work upon operations one style at a time.

STYLE PERFORMANCE EFFICIENCY

Style performance efficiency is the average performance experience with respect to standard achieved in producing specific styles. Normally each style represents a fabric, a color and a pattern. It is a useful value to have when comparing work requirements for a style mix to available plant capacity. Different fabrics may affect operation efficiency. The impact of fabrics upon costs is a valuable costing tool This calculation is not frequently used because it is difficult to obtain the data to compute it, and because few knew how to make the computation correctly.

Computer systems can collect the data easily. Alternate ways to compute style performance efficiency follow.

As operators' time spent on each operation is recorded along with the quantity produced, it becomes possible to compute both the operation efficiency and the style efficiency.

Two ways to compute the weighted average performance efficiency for a style are covered in the following.

As a complete style is produced, work flows through a complete operational sequence. Planning effort is made to maintain the flow of work by assigning enough operators at each operation. The efficiency realized at each operation is a composite of the efforts of several operations.

Operator-days of work at each operation serves as a weight. This weight governs the influence of each operation upon the performance efficiency for the style as a whole. Any production rate will serve for computing the operator-days to apply at each operation, for present purposes, because it is only significant that there be some proportional number of operator-days at each, providing for a balanced flow of work.

The operator-day requirements at any operation for any given production rate could be computed by means of the following formula, assuming standard performance:

Operator-day requirements at standard equal:

$$\frac{\text{Desired 100s/day X SAH/100}}{\text{Hours Worked per Day}}$$

To compute operator-day requirements for an operational efficiency other than 100%, it is necessary to divide the value at standard by other performance efficiency. The expression for doing this may be factored into the following parts:

Operator-day requirements at actual efficiency equal:

$$\frac{\text{Desired 100s/day X SAH/100}}{\text{Hours Worked per Day}} \times \frac{1}{\text{Opr. Perf. Eff.}}$$

It may be seen that the operator-day requirements at standard would be natural weights for a harmonic mean of the operation efficiencies. A weighted harmonic mean would be computed like this:

$$\frac{1}{\bar{e}} = \frac{1}{n} \times \text{Sum of } \frac{w_1}{e_1} + \frac{w_2}{e_2} + \ldots\ldots + \frac{w_n}{e_n}$$

Symbolism:

$\bar{e}$ is average efficiency

e_2 is performed efficiently at a certain operation

n is total number operator-days (total of operation weights, total operator-days applied per day to style.)

w_2 is the weight at the second operation (standard operator-days/day)

An example follows. The operation list is presented below for Widgit X. With the time standards expressed in hours per hundred, the daily production rate will be 1,000 units per day or ten hundreds per day. The work day contains eight hours. The operator-day requirements at standard performance to achieve this rate are computed for each operation. The performance efficiency ratios actually experienced in producing Widgit X have been taken from an operation efficiency file and set down in the adjacent column.

When the operator-day requirements at standard are divided by the operation efficiency achieved at the operation, the result is the operator-day requirement corrected for actual efficiency. This must be done for each operation. Then, the two operator-day columns are summed up.

The numbers for Widgit X are listed in Table 12.2.

When the total hours earned and clocked producing a certain quantity of a style are accumulated, a correct weighted average style efficiency may be obtained by dividing the total hours earned by the total clocked. Load hours would be the same as clocked hours, it viewed looking toward the past. Hence, the average would be:

$$\text{Style efficiency} = \frac{40.17}{43.627} = 0.9208$$

(may be called 92.08%)

TABLE 12.2
CALCULATING OPERATOR REQUIREMENTS

Opn. No.	SAH /100	100s /Day	Std. Hrs. /1000
1	0.500	16	5.00
2	1.000	8	10.00
3	0.250	32	2.50
4	1.600	5	16.00
5	0.667	12	6.67
Totals:	4.017		40.17

Std. PO-RQ	Opn. Eff.	Load Hrs /1000	Actual OP-RQ
0.6250	1.15	4.348	0.5435
1.2500	0.80	12.500	1.5625
0.3125	1.25	2.000	0.2500
2.0000	0.85	18.824	2.3529
0.8333	1.12	5.955	0.7440
5.0208	5.17	43.627	5.4529

Arithmetic Mean: 1.034

A straight arithmetic mean of the operation efficiencies, provides a wrong and misleading result.

$$\text{Arith. Mean} = \frac{\text{Sum of Effs}}{\text{No. of values}} = \frac{5.17}{5} = 1.034$$

This arithmetic mean differs from the weighted average obtained dividing total hours earned by the total hours clocked, 0.9208.

If these total hours had not been accumulated, but the average operation efficiencies were individually known, it would be possible to compute a weighted

Weighted Harmonic Mean:

$$\frac{1}{e} = \frac{1}{5.0208} \text{ X Sum of } \frac{0.625}{1.15}$$

$$+ \ . \ . \ . + \frac{0.8333}{1.12} = \frac{5.4529}{5.0208}$$

$$\bar{e} = \frac{5.0208}{5.4529} = 0.9208$$

= weighted harmonic mean performance

This weighted harmonic mean performance is correct where the ordinary arithmetic mean is not. The operations with the largest time standards were the ones done least efficiently in this case. These operations took the most time and involved the most people. Therefore, they had more effect upon the overall efficiency for the style and caused it to be lower than standard, 0.9208.

As a second example, use the same Widgit X with rearranged performance efficiency values. This time, the operations taking longer times are performed efficiently and the ones with shorter times are done inefficiently.

TABLE 12.3
RECALCULATING OPERATOR REQUIREMENTS

Opn. No.	SAH /100	100s /Day	Std PO-RQ	Opn. EFF.	Actual OP-RQ
1	0.500	16	0.6250	0.80	0.7813
2	1.000	8	1.2500	1.15	1.0869
3	0.250	32	0.3125	0.85	0.3676
4	1.600	5	2.0000	1.25	1.6000
5	0.667	12	0.8333	1.12	0.7440
Totals:	4.017		5.0208	5.17	4.5798
		Arithmetic Mean:		1.034	

Harmonic Mean:

$$\frac{1}{e} = \frac{1}{5.0208} \text{ X Sum of } \frac{0.625}{0.80} + \ldots + \frac{0.8333}{1.12} = \frac{4.5798}{5.0208}$$

$$\overline{e} = \frac{5.0208}{4.5798} = 1.0963$$

= harmonic mean performance

In this case, more time is spent working efficiently and the weighted harmonic mean correctly assesses the impact upon the overall performance. The result is better than the previous example at 1.0963.

The straight arithmetic mean remains constant in both cases. It does not reflect what has happened. If a weighted arithmetic mean had been computed multiplying the standard operator requirements by the operation efficiency values, summing them and dividing by the total operator-days, the results would still have been wrong.

There are only two correct approaches:

a. Accumulate and divide the total hours earned by total hours clocked producing a style quantity.

b. Use the individual operation efficiencies and compute a weighted average harmonic mean.

Either approach or both can be derived easily with a computer program.

OPERATOR AND OPERATION CROSS REFERENCE FILE

A cross-reference file between operations and operators should be established. A program can be written which makes it possible to enter an operation number and find the operators who can perform it. Another program should allow one to enter an operator's name or number and discover the operations which that person is able to perform, together with the operator efficiency for each operator.

This cross-reference file is necessary for sound operator assignment. It makes it possible to find the operators who can be assigned to a given operation. It also makes it possible to select the operators best able to do the operation first.

The operator's overall efficiency is the weighted average of various operation performances. When selecting an operator to assign, it is desirable to know how well an operator is able to perform each different operation. To evaluate this, it is necessary to know how many hours of experience the operator acquired in achieving the performance level. Therefore, the basic file should contain the hours on the operation, the operator efficiency and the last date on which the operator performed the operation.

ABSENTEEISM

Absenteeism upsets production plans and schedules. Absenteeism is attacked on two levels:

a. Influencing personnel to attend work.

b. Weighing production plans to compensate for absenteeism which will occur.

Most plants prepare an absence and lateness report daily. It identifies those who are late or absent. It helps supervisors to influence workers to attend regularly and be on time.

Normally, the plant personnel clerk prepares an analysis of absenteeism. This report establishes the current rate of absenteeism as a percentage of gross capacity hours daily. It does this by work center and by plant. This current rate is used by supervisors to follow up on absentees. It is used by plant management to check on supervisor effectiveness.

The analysis may also provide a 20-day moving average rate of absenteeism by work center by supervisor and by plant. This rate is used to reduce gross plant capacity to available capacity.

LATENESS AND ABSENCE REPORTS

When the payroll gum sheets are scanned after the first two hours under a two hours Update System, the operators turning in sheets are identified. The employees listed on file who are not identified at this time are recorded absent, at least for a quarter of a day.

The second coupon or the time clock stamp, after the operator-identifying coupon on the worksheet, gives the starting time. Any employee coming in late has obtained a starting time coupon or clock-in from the work center or line supervisor bearing the true starting time. If this time read-in by the OCR scan is later than the appointed starting time, then the employee is recorded late by the actually differing time interval.

Some firms record reasons for absences and tardiness and distinguish between excused and unexcused ones. Other firms consider that all absences are inexcusable and that only a certain limiting number of absences can be tolerated in a time period before discharging the employee.

If excuses are accepted, there should be an input of the excuse when accepted. Any later report would state the reason beside the excused absence, usually by a code number or letter.

If excuses are not accepted, then they are not entered. Only the points at which verbal and written warnings are given would be reported. The signals to do this would be given to supervisors automatically from the report produced.

It is recommended, in the interest of fairness and simplicity, that a system be adopted in which absences and tardinesses are never excused but simply limited to certain numbers of occurrences. After a period of time during which the employee has a clear record, the report programs should be able to clean infractions from the record. The lateness and absence report in such cases would be produced daily. It should be delivered to work center supervisors.

a. The plant.

b. The work center.

c. Date.

d. Names absent.

e. Number of quarter-days absent.

f. Type of warning due, if any, for absence.

g. Names of those late.

h. Decimal hours of lateness (always less than 2 hours).

i. Type warning due, if any, for tardiness.

j. Total time lost.

   (1) From absences.

   (2) From tardiness.

k. Total at bottom of page of time lost from absence and lateness.

The time lost is the sum of the hours and the quarter-days of absence plus the total decimal hours of tardiness. These daily totals can be summed to weekly totals. These weekly totals may be used in the absenteeism analysis.

Absence or tardiness on an occasion for overtime may be recorded separately on the report, but would not enter the weekly total. Supervisors and plant manager may have an interest in this type of absence for disciplinary reasons. Overtime is not included in the normal capacity hours for a plant. For this reason, absence from overtime periods is not normally considered in the absenteeism analysis where the percentage of normal capacity lost due to absenteeism is computed. The weekly total is used in this computation and so should not include absence from overtime.

ABSENTEEISM ANALYSIS

This analysis is intended to serve two ends:

a. To evaluate current absenteeism rates in the work center and plant.

b. To determine the expected loss of plant capacity due to absenteeism.

Absenteeism Analysis normally is reviewed weekly. The plant manager and all supervisors should receive and review the report.

The analysis report may be by work center or by production line. The plant as a whole is usually totaled. The data are taken from the Lateness and Absence Report. Total hours lost due to absenteeism are divided by gross operator-hours of capacity to give the percent of capacity lost during the past week.

The analysis report also carried forward a 4-week moving average absenteeism loss for each work center or production line allowing the plant to compare with the last week's results.

The capacity loss due to absenteeism is used by a plant manager to evaluate the impact of absenteeism upon his plant. The analysis can help the manager to judge the supervisor's influence upon absenteeism in the work centers or production lines.

The moving average absenteeism losses for the work centers, production lines and the plant are percentages which can be used to reduce gross capacity to available capacity. The absenteeism loss is a percent of gross operator-hours to deduct as unusable. This loss may be converted to the number of utility or standby operators needed to maintain full plant operation.

The method to determine the gross plant capacity in operator hours involves these steps:

a. Determine the number of employees of record.

b. Determine the number on leave of absence (sabbatical, sick leave, etc.).

c. Determine the number of employees on vacation for the period of interest.

d. Determine the number of employees laid off for the period of interest.

e. Sum the numbers from 2, 3, and 4 above and subtract this from 1 above to get the number able to work.

f. Determine the daily and weekly working hours.

g. Multiply the working hours per time period (day or week) by the number of people able to work to get the gross operator hours of capacity.

The expected loss of hours owing to absenteeism must be removed from gross operator-hours before comparing work requirements for style batches to the available capacity.

OFF-STANDARD ACTIVITIES AND EXCESSES

This section provides definitions of off-standard activities and excesses. It explains which losses of time have the effect of reducing the direct labor capacity of the plants. It explains how the gross capacity is reduced to available capacity before the work requirements are compared to it.

OFF-STANDARD ACTIVITIES

Off-standard activities are those which are performed without comparison to any standard. The time is paid for without measurement of results. Off-standard activities constitute a loss of direct labor time which reduces direct labor capacity.

Because off-standard activities all have this common characteristic, they are treated separately from other things that also create excess costs. Only certain off-standard activities must be defined and classified by type.

The following categories represent typical definitions of off-standartd activity and pay policy:

Waiting Time

An incentive employee may wait on work instructions, supplies, or findings. The waiting time shall be verified by supervisor. Waiting time is paid at the average hourly rate.

Machine Delay

This occurs when the machine breaks down and the operator must wait for a mechanic to repair it. A machine delay is paid at the average hourly rate.

Repair for Others

When the operator does repair work originally done by another operator, the operator is entitled to repair time. (If the operator did the bad work, the operator would repair his or her own defective work during on-standard time.) This will be paid at average hourly earning.

Loss from Faulty Cutting

This occurs when bad cutting work causes a delay in the operation. The time lost is considered to be that which is in excess of the standard time for the work. This condition must be verified and certified by the supervisor. This loss would be paid at the operator's average hourly rate. Coupons would be taken for doing this work. The excess time is defined as the difference between the actual time and the standard time.

Loss from Defective Findings

This time is lost when labels, linings, buttons, tapes, zippers or other such findings are defective and cause the work to be slowed. This condition again must be verified with the supervisor. When it occurs it is paid at the average hourly rate. Coupons would be taken for doing this work. The excess time is defined as the difference between the actual time and the standard time.

Making Samples

This excess occurs when the operator is given samples or prototypes to make. This work would have no time standards. It would be assigned at the convenience of the company. It would be paid at the average hourly rate the operator had earned.

Taking Inventory

This work would only be assigned during inventory-taking time. Operators would be assigned at the company's convenience and would be paid at the operator's average hourly rate, or at some hourly rate pre-established by the firm (sometimes, in union negotiations).

Service Charged to Buyer

This would be a special service (i.e., an extra label or a change of labels) for which the buyer has authorized a charge. This service would be performed at company convenience. The operator would be paid at average hourly wage.

Personnel Conference

This time away from work represents time spent by a piecework operator who is called away from the workplace for an interview or a discussion with management. It also comprises time spent at company authorized meetings. This would be paid at average hourly rate.

Time Work

These time work jobs are temporarily authorized and paid at an hourly rate which depends upon the circumstances at the time, or the management policy established.

a. Inspect the work of others.

b. Bundle handling.

c. Utility work.

d. Expediting a production order in process.

e. Clerical work.

Unmeasured Work

In this category the job being performed is a production job with a measurable result and for which no time standard has been established, but for which it is expected that one will be set. Unmeasured work is authorized by the supervisor. The operator will clip the appropriate piecework coupon without a time standard. The operator will be paid at average hourly rate. The amount of unmeasured work is often used to evaluate the effectiveness of the plant or division engineer.

Set-up Time Delay

A delay which involves both machine delay and waiting time. It is associated with getting a workplace set up properly to perform the operation. The time must be verified by the supervisor. It would be paid for at average hourly rate.

If a direct labor operator is assigned timework, the hourly rate of pay depends upon the circumstances defined. Hourly rates which may be used fall into three classes.

Transfer at Average

In this case the transfer occurs for the convenience of the company and the operator is paid 13-week average hourly rate, or the value of the actual coupons earned during that time, if any, whichever is greater.

Transfer at a Percent of Average

The transfer occurs at the convenience of the operator owing to some mishap such as machine breakdown. The full 13-week average rate is reduced by a specified percentage, but remains higher than minimum.

a. An hourly wage which is set as some percent of average may be established by the management of the group or division. It could be any value between minimum and average.

b. Union contracts often specify an hourly rate at the midpoint between minimum and average for cases in which it may be as much in the company's interest as in the worker's interest to transfer to the job. This corresponds to a specific percent of average.

Transfer at Minimum

The operator is given work for their convenience rather than that of the company. In this case the minimum hourly wage is used.

Transfer at Special Incentive

In this case coupons will be clipped and the pay will be determined by an established incentive plan. The operator will be paid an incentive rate based on specified percentage of the base rate plus another specified percentage of the value of the coupons earned.

Excesses

Conditions which create costs in excess of direct labor cost are called excesses. These costs are defined here.

Off-standard activities create excess costs and at the same time reduce direct labor availability. They have been separated from this category, leaving those excesses which need to be accounted for but which do not directly reduce applicable direct labor.

Some excesses do affect the performance efficiency of direct labor. This is subject to measurement and is treated differently than off-standard losses. Such excesses are included here. They are identified as:

a. Bonuses.

b. Overtime premiums.

c. Make-up to minimum hourly wage.

d. Make-up to average hourly wage.

e. Trainee make-up.

f. Paid holidays.

g. Paid vacation.

h. Miscellaneous excesses.

OFF-STANDARD TIME LOSS REPORT

Each activity to be recognized as off-standard time is identified in another section. Off-standard time is to be recorded on each operator's gum sheet and verified by the supervisor when it occurs.

When the operator's work sheet is scanned by the OCR wand, the off-standard data are put into the payroll system. The pay for off-standard time is computed and then the increments of time are accumulated in the appropriate off-standard categories. The total of each kind of off-standard time is accumulated each week. The final totals will be complete by the weekend for two different analyses, namely:

a. Labor time and cost analysis.

b. Off-standard time loss report.

The off-standard time loss represents the use of direct labor for nonproductive activities. This activity constitutes a measurable deduction from plant capacity. The data can be analyzed by production line or work center or the whole plant can be determined by the same method. The operators would have to be identified correctly with their work centers when work center capacity is defined.

The total operator-hours lost on off-standard activity are divided by the gross capacity to obtain a fraction or a percent of gross capacity lost. This would be the last week's rate of loss.

The weekly rates of loss by work center and plant should be entered into a 4-week moving average. This is a more stable average which can be used to compare with the last week's value. Also, it can be used as a basis for reducing plant capacity owing to the loss of time in off-standard activities.

This off-standard time analysis corresponds in format to the absenteeism analysis. It has the last week's percentages to compare to the 4-week moving average, showing them to be better or worse than average.

FIXED TIME LOSSES

The fixed time loss is the last of the three losses that reduce gross capacity to available capacity. They are as follows:

a. Absenteeism loss.

b. Off-standard time loss.

c. Fixed loss.

Fixed losses are those that occur every day for every operator in the same amount. The maximum possible time passed by an operator in the plant for a full work day may be 480 minutes or 8 hours. This must be reduced for fixed losses of capacity that routinely occur as a matter of policy.

Fixed losses might be incurred as a result of break periods. If there were two 15-minute break periods, then 1/2 hour per day would not be used for production. If five minutes at the end of the day were set aside for work place clean-up, then these five minutes would not be used for production. Such losses as these are fixed losses.

The amount of time in the fixed losses each day should be summed up. This total would be divided by the time in the work day to obtain a fraction or percentage which would represent the time lost. It would be applied as a system constant to the plant.

This value would remain constant so long as the fixed loss policy of the organization remains the same. The maximum useful time in a work day or work week would not include the constant fixed losses.

Engineers will be quick to point out that, at least in some plants, the time standards are computed using allowances for personal time, fatigue, and ordinary delays which take breaks and clean-up periods into account. Perhaps this is true in some places and is simply assumed to be so in others. It can only be asserted here that the matter is not uniformly dealt with everywhere in industry.

The PF&D allowance should be analyzed into its detailed parts. It should be inspected to see that each element is adequately allowed for. It should be clearly determined that break periods, clean-up time, and the like either are or are not included. When they definitely are included in this allowance, they do not need to be treated again as fixed losses of capacity. When they are not, they do need to be.

LABOR COSTS

Labor costs will be analyzed and accumulated for: 1) management reports, 2) legal records, and 3) financial reports. The different files which must be maintained for these purposes will now be reviewed.

It is proposed to read coupons for production reports into the system every two hours or daily to provide work-in-process reports on a nearly current basis, operating in batch mode. From this input, data can be accumulated to process the source-to-gross payroll once daily. Labor costs subsequently may be accumulated in the different files. This would be a part of the source-to-gross segment of the payroll programming.

DIRECT LABOR

Persons performing piecework and timework operations which contribute directly to the production of the products are classified as direct labor. When they happened to be assigned indirect labor jobs, it must be specifically declared on the worksheet bearing the piecework coupons.

The source-to-gross payroll procedure normally provides for the following assemblies of data:

a. A payroll journal (individual operator records).

b. Labor cost analysis by category, work center, department.

c. Thirteen week moving average hourly pay

    (1) Currently accumulated.

    (2) Distribution of employees over average hourly rates of pay.

This can be done for overall average rates as well as for specific incentive performance earning hourly rates.

The payroll journal is a file which provides the basis for an individual labor cost analysis. Operator names may be placed in alphabetical order in this file but must be cross-referenced down to their work centers and operations.

Reports of individual labor analysis will be desired, organized by:

a. Plant.

b. Department.

c. Work center.

d. Operation.

e. Operator.

f. The payroll week identified by ending date.

g. Date of the day.

h. Name of the day worked.

Lines beside operator names on an individual labor analysis would include data under the following column headings:

a. Employee clock number.

b. Employee's name.

c. Hours clocked on the operation.

d. Hours earned on the operation.

e. Dollars earned on the operation.

f. Performance efficiency on the operation.

g. Operation category (regular job, transfer, at average, etc.).

h. Totals for the operator day.

(1) Total hours clocked.

(2) Total hours earned.

(3) Total overtime hours.

(4) Dollars not earned on the operation.

(5) Dollars earned by the operator.

(6) Off-standard hours.

INDIRECT LABOR

Persons performing jobs which provide supporting services but which do not directly cause production are classified as indirect labor. There are no piecework earnings calculated in this category. All indirect employees are paid by the hour.

Indirect labor categories such as the following will be established for these employees:

a. Supervision.

b. Mechanics.

c. Bundle handlers or toters (sometimes classed DIRECT).

d. Quality control inspectors.

e. Clerical.

f. Maintenance.

g. Administrative, which includes:

(1) Production control.

(2) Data processing people.

(3) Others.

Indirect workers will have their hourly earnings posted in a different section of the payroll journal from direct labor employees.

An individual indirect labor analysis should be prepared by plant and by payroll week listing employees in each category. It should give actual number of employees, the total dollars paid in the week of the report, actual overtime dollars paid in the week, or budgeted number of employees, budgeted dollars for indirect pay, the actual variance from budget. This report may be provided to Plant, Division and Group management.

STATISTICS

Statistics and analyzed data will be needed for insights and interpretation of the performance of direct labor and of operating costs. Reports would be made routinely by week. A daily inquiry capability should be provided for plant managers and supervisors.

Data and statistics such as these should be compiled for direct labor:

a. Hours earned.

   (1) Total hours earned.

   (2) Hours on standard jobs (transfer, split incentives, etc.).

   (3) On piece rates.

b. Hours spent.

   (1) Total hours.

   (2) Off-standard.

   (3) On piecework.

c. Earnings.

   (1) Average earnings on piecework.

   (2) Average hourly earnings on hourly paid activities.

   (3) Overall average hourly pay.

d. Performance.

   (1) On piecework.

   (2) Overall performance efficiency for the period (including off-standard time).

e. Cost of production.

   (1) By plant.

(2) By department.

(3) By work center.

(4) Sequence level.

f. General statistics by department and work center.

(1) Current average hourly earning.

(2) Current average performance efficiency.

(3) Current average overall efficiency.

(4) Thirteen week moving average performance efficiencies to compare with the foregoing.

(5) Total units produced.

(6) Direct labor cost per unit.

(7) Indirect labor cost per unit.

(8) Total labor cost per unit.

(9) Number of operators over 100%.

(10) Number of operators in make-up.

(11) Total payroll makeup by work center.

LABOR COST ANALYSIS

An analysis of labor costs would be prepared for both direct and indirect labor for each plant. The direct labor costs would be broken down by department and work center within the plant. A report of the analysis would be prepared weekly.

Apparel plant departments normally fall under one of the following names:

a. Cutting.

b. Sewing.

c. Finishing/Pressing.

d. Warehousing

   (1) Materials.

   (2) Finished Goods.

e. Shipping.

Departments of other plants may have names such as these:

a. Garnetting.

b. Knitting.

c. Textiles Finishing (dyeing, coating, etc.).

d. Automatics.

e. Tenter Frame.

f. Bonding.

g. Quilting.

j. Metal Working.

i. Batting, fiber-filling, or down-filling.

Departments may be broken down into production lines or work centers. Where the work centers are large enough to be significant, the departmental analysis may be further broken down within a production unit.

Each direct labor analysis will comprise the following items:

a. Direct Labor Costs:

(1) Piecework.

(2) Timework.

(a) Unmeasured Work.

(b) Other.

(3) Sub-total.

b. Excess Costs.

(1) Payroll Make-up.

(a) Experienced Operators.

(i) Transferred at average.

(ii) Transferred at a percent of average.

(iii) Transferred on a split incentive plan.

(iv) Transferred at minimum.

(2) Off-Standard Activities.

(a) Machine downtime.

(b) Waiting for work or instructions.

(c) Meetings and interviews, authorized or required.

(d) Other off-standard activities.

(3) Overtime.

(4) Other excesses.

(5) Sub-total.

c. Other direct costs.

(1) Holiday.

(2) Vacation.

(3) Miscellaneous.

(4) Sub-total.

d. Departmental Total Direct Labor Costs.

e. Transfer Wages Paid During Week.

(1) Transferred at average.

(2) Transferred at a percent of average.

(3) Transferred on a split incentive.

(4) Transferred at minimum.

The wages paid to transferred operators are included in the departmental total direct labor. They are extracted and listed last as information.

The indirect labor for the whole plant may be reported together all at once. All indirect labor is paid as timework. The following items are included in the indirect labor cost analysis:

a. Supervision.

b. Mechanics.

c. Bundle handlers.

d. Quality control inspectors.

e. Clerical.

f. Maintenance and housekeeping.

g. Administrative

(1) Production control.

(2) Data processing.

(3) Other functions.

Beside all of these direct and indirect labor items, on the same line, the following data would be presented in columns going from left to right:

a. The number of employees involved.

b. Dollars accumulated in the current period.

c. Percent compared to total direct labor.

d. Percent goal.

e. Dollars accumulated year-to-date or month-to-date or week-to-date as specified on the report.

f. Accumulated percent compared to direct labor on the year-to-date or specified basis.

g. Last year's equivalent period date.

h. Percentage of last year's equivalent period.

A report of this analysis would be provided to Plant Division or Group management. It could also be given to supervisors and to the engineering department.

PRODUCTION STATUS

Knowing the status of production enables control over the movement of work-in-process. This makes it possible to:

a. Satisfy customer deliver expectations.

b. Make fuller use of the operator hours capacity of the plant.

c. Avoid accumulating too much inventory in the process.

In order to realize these benefits it is necessary to acquire and to process a large amount of data. In the past, it was very difficult to do this manually. Computerized payroll procedures make it possible now to deal with the data in timely fashion. The reports derived do not require as much clerical or supervisory time as before.

Detailed records of the positions of all bundles in process are available from the piecework payroll reports. Specific bundle quantities are identified with certain styles and production orders. Sizes and colors are also known because the coupons relate to a bundle list made for the production order.

The frequency with which reports of the coupons are made governs the exactness with which bundles may be located along their routes-in-process. Real time reporting affords the maximum exactitude. Reporting after intervals of time affords approximations which are less accurate as the intervals grow longer.

Several payroll systems now pick up data at the operator's machine as the work flows. However, a two-hour reporting interval provides sufficiently close approximations of bundle locations for good control of the work flow. In many firms a daily run is adequate. Batch processing is usually less expensive to operate than an on-line system.

In fixed production arrangements that continually have the same sequence of operation for all styles, it is feasible to select key operations at chosen places or at points of convergence to be control points for reporting purposes. It is possible to keep track of the work flow by reporting the production only at such key operations, in the appropriate circumstances, and not reporting production from all the operations.

The previous chapter describes a suggested definition of complex work flow routes by sectors. The sectors are defined by sub-assembly route and sequence level. A bundle is considered produced out of a sector when all the operations in the sector are done for it. Operations within a sector could be performed in any sequence. Each sector would be regarded as a key point.

There are five significant aspects of the production status which should be observed and evaluated frequently. These should be viewed by sequence level, work center and plant as a whole. They are as follows:

a. Production (output).

b. Inventory in process.

c. Bundles bypassed.

d. Operations performed out of sequence.

e. Average inventory per operator relative to production.

These items were covered in the preceding chapter.

COST OF PRODUCING STYLES

Merchandising and manufacturing managements have a need to know the real costs of producing different products. The information can be made available upon the basis of completed styles. It would include any or all of these for each style:

a. Total completed style costs.

b. Total completed style operator-hours.

c. Total units produced.

d. Cost in dollars per hundred.

e. Cost in operator-hours per hundred.

f. Dollars per operator-hour spent.

Although the data would be accumulated on a daily and weekly basis, the final results may comprise data acquired during several weeks, spanning payroll periods.

It is possible to obtain direct labor costs of producing different styles from the piecework payroll. Indirect labor costs are available for weekly periods. Material costs can be computed following usage reports.

Sorting out and accumulating labor and material costs per style are very ordinary data processing tasks when the costs are already available. As data are put in, they can be accumulated in appropriately identified registers until the style quantity being produced is complete.

Procedures to accomplish this end could be incorporated in computer sub-routines performed as soon as the data becomes available. Such procedures will be outlined here.

ASSEMBLING DIRECT LABOR COSTS BY STYLE

At the time the coupons are read, the style, time standard, and quantities are discovered and associated with an interval of time during which the work was really done. It is possible to determine immediately the actual labor costs, the earned hour rate, and the performance efficiency for the style.

The direct labor costs from the piecework payroll are analyzed by style when put in. Accumulating the costs by style is simply a matter of establishing registers in which to accumulate the costs. Both the operator-hours actually spent and the costs incurred should be accumulated.

A payroll sub-routine can be provided to establish registers for each new style and production order during data input. These would accumulate the increments as entered into the plant. Cost totals can be accumulated by production order. This allows costs to be captured for completed style costs, and allows the separation of costs from uncompleted style costs.

Total costs in operator-hours and in dollars can be respectively summed and carried forward as production orders are completed. The units produced can be accumulated from the production orders themselves, using the quantity ordered or can be obtained from completion reports, using the units really completed. Costs and quantities associated with yet uncompleted production orders would not be included.

Average direct labor costs for each style may be readily computed, using the accumulated totals from completed production orders. These may be expressed as dollars per 100, operator-hours per 100, and dollars per operator-hour. It may be desirable to report these actual averages in comparison to the standard cost values.

COST PRORATED TO STYLES

Both the indirect labor and the excess costs of direct labor can be prorated to production orders each week. The amounts prorated may later be added for the completed orders of the same style. This action separates the completed style costs from the uncompleted style costs.

Indirect labor costs can be prorated weekly to production orders based on the operator-hours actually spent on each one. When a production order is completed, no further coupons or hours actually spent are reported for it. No more indirect labor would be associated with an order than is proportional to the direct labor time really spent. This relates indirect labor to styles. When each production order is completed, the increments of indirect labor from previous weeks are totaled. Actual hours and dollars from different production orders for the same style can then be summed.

The quantities of production orders are matters of record. They too can be summed by style. Hence, it is possible to compute average indirect labor costs expressed as dollars per hundred, operator-hours per hundred, and dollars per operator-hour. These are associated with the same production orders and styles as the direct labor values. Because of this, the average indirect labor may be added to the direct to give a total average labor costs.

Payroll make-up for trainees and experienced operators who turn in coupons is immediately associated with the production order by style. Payroll makeup can be accumulated like direct labor costs. It is an excess cost which would not have to be prorated.

Other excesses which result from time lost on off-standard activities would not be identified with particular styles and production orders. Consequently, it would be necessary to prorate these costs the same way as indirect labor. Each week the total off-standard excesses would be prorated to each production order in proportion to the direct labor hours spent on each one.

The effect of absenteeism is to increase the amount of off-standard time lost waiting for work or instructions. This activity can remain a prorated excess. Operators may be freely assigned to one style or another. For this reason, it would be difficult to say which style in process may be most affected by any particular absence. It would not be impossible because it is a matter of record as to which styles are in process. It is also a matter of record which styles an absent operator might have worked on if she or he were present. The programming and the computer running time required to relate absenteeism to particular styles may be more expensive than the value of this warrants. Prorating will probably be sufficient in the long run.

Start-up costs which are not captured as experienced operator make-up or trainee make-up might not be correctly associated with the production order and style unless a specific declaration is made of it. Such costs would be related to mechanical set-ups, equipment rentals, or hired services, such as those of an electrical contractor, or a manufacturer's service call. All charges to the company should be fully identified by the person authorizing them. The authorizer should relate the service to the style and production order necessitating the service.

Mechanics who make those equipment set-ups required in order to produce a style can use the same job-clocks as the operators to record the time they spend doing this. This will serve to report the time and cost per style and production order specifically. Also it will eliminate this time from the indirect labor total which is distributed proportionately to all production orders running during the week.

Operators who wait for mechanics to make set-ups should identify the style and production order for which they wait when they clock their waiting time.

The excesses prorated to production orders may then be summed by style and reduced to averages per hundred like the indirect labor was. These would be associated with the same production orders and styles as the indirect labor. Therefore, it would be appropriate to add these average off-standard costs per hundred to the other corresponding average costs per hundred.

Factory burden consisting of rent, electricity, water, sewerage, and such may be taken into account at some point. It is not mandatory that these costs be prorated to production orders and then to styles, but it can be done. It is one way to consider these costs but other options might be chosen as well. These might just as well be summed and applied as the same average value per hundred items produced.

MATERIAL COSTS PER STYLE

Material usage is prescribed by production order. A Bill of Materials is extended to specify appropriate quantities of each item of material. The costs of these are associated with the production order and style.

In accumulating true style costs, it is better to obtain reports of the actual usages than to use the predicted quantities of material. It is always possible that some condition or circumstance may cause the real usage to differ from what was expected.

Most reports would be associated with particular production orders, so the usage costs could be accumulated by style. Some items, such as thread or buttons, may be difficult or impossible to account for by style. It may be necessary in such cases to use standard values for the usage of each item by style. It would be more desirable to use such an estimate than not to account for such items at all.

HIGH AND LOW OPERATOR PERFORMANCE REPORT

Some operators may earn exceptionally high amounts from piecework and others may remain continually in make-up. These circumstances should be investigated.

When there are many cases of extreme earnings, high or low, there is reason to investigate as to: Is the incentive system deteriorating? Is the system effective? Are methods changing without proper recognition? Are time standards properly related to the cyclic method or to auxiliary activities?

The reason for monitoring high and low performance is to make sure the incentive system is truly fair and equitable for all participants. Definition of what is high will always take place.

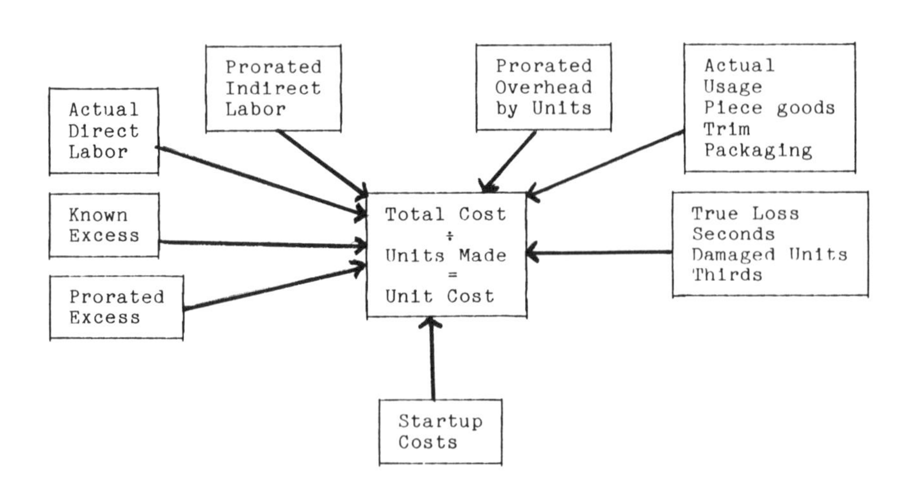

FIGURE 12.1. TRUE STYLE COSTS

Either the workers will achieve as much as they think management will endure paying for and limit their productivity at that point, or the management will set the limit, establishing a proper standard for performance. Standards should reward people for attainable productivity, and encourage it to be achieved.

An exception report is needed to reveal operators who achieved earnings and performance efficiencies at particular operations which are above or below desirable limits. Performance outside certain bounds gives indication that something unusual pertains to the situation. Difficulties or indolence, which are out of the ordinary, may bring the productivity down. Some method change, improper time standard, or extraordinary performance may bring productivity up. In any case it is necessary to learn the reasons why a condition of exceptionally high or low performance exists.

Engineers sometimes claim an operator is super-skilled in an effort to justify the loose standards they have set. Method time analysis affords an objective means of determining whether or not the movements being really used are defined in the operational method or not. If an operator is discovered to have an exceptional ability to performm the clearly defined method, then there exists no reason to change a time standard. When different motions are involved or when auxiliary operations have been dispensed with, reasons for change may exist.

A <u>High-Low Performance Report</u> names the operator, operation, and work center of the plant in which an exception occurred. The report is designed for supervisors, engineers, and plant managers. Normally it is produced weekly, corresponding to the weekly payroll. It can be prepared for longer or shorter periods if needed.

The piecework performance of operators on operations where they performed exceptionally is reported in terms of performance efficiency and dollar earnings. Overall operator earnings and average efficiency also are given, together with total off-standard activity times. Unwarranted shifts of working time into off-standard time could cause exceptional performance to appear.

Discovery of low performance may precipitate three kinds of action. First, undersirable or improper working conditions may be corrected. Second, attention might be given to the operator's training and motivation. Third, another method analysis might be made to ascertain that the method used and time standard are compatible.

Discovery of high performance leads to different action and deals with personnel problems of another class. An operator achieving 140% of an operational standard may be doing so through skill and effort. Where there is an incentive allowance of 12.5%, this could be reported as 153% of standard. There is a good chance, anyway, that this owes more to the ability and application of the operator than to laxity of engineers setting standards. When operators are making 175% to 225% of standard or higher, there is a reasonable indication the the engineering standard is not appropriate for the work being done.

Upper boundaries should be set after some study. All operators who exceed the upper boundaries should be reported to plant management and supervision, but no reproach or alarm should be expressed to the operators involved. The report should be made in order that the circumstances might be investigated. They should be objectively investigated.

No industrial engineer has such transcendent skill that he will always design methods that operators cannot improve upon. Although the time standards for his methods may be set by MTM-1 or 4M techniques, so as to be exactly appropriate to the motions intended for performance, some operator may sooner or later discover an improved method for doing the job. The operator will begin to run away with the time standard which was really set for a slower combination of movements.

The fact will require restudy. The operator may become fearful of loss of pay if the full potential of the more productive method is exercised and may ease up the work pace. An operator may fear that the management will cut the standard to get more production at the same cost, and that there will be no personal benefit from it. Peer pressure might also be applied by those other operators who do not wish to be made to appear slow in comparison to this operator with the better method.

If productivity is to be improved, some way must be found to circumvent these fears and the foundations of these fears. Productivity will not be improved if adversary attitudes are preserved between labor and management. This is sometimes achieved by awarding a bonus to the innovative operator.

Of course, it is also possible that the time standard was not correctly set in the first place. The motion analysis may not have been correctly performed. An MTM-1, 4M or other accepted standard could be established in error. Then the tribulation of resolving the error between operator and engineer begins.

This trouble is not likely to occur as frequently with MTM standards as with those established by time study. Methods are normally better defined with MTM studies than with time studies. The predetermined times comprise a uniformly applied performance leveling system, while time studies are individually and inconsistently rated for a normal level.

Most firms have many old time standards established by stopwatch time study for operations which continue to be performed on both and old and new styles. Some have been set either judgmentally or through personal negotiations, and no documentation for them can be found.

Several approaches to change may be considered. Briefly stated they are:

a. Break the original operation into different parts, changing the method, setting the time standard by 4M, and retraining operators.

b. Recognize that an error was made in setting the original standards and make a time-limited monetary settlement with the operators involved for agreement to work against corrected 4M standards in the future.

c. Recognize that the operator has discovered a better method; remunerate the operator on the basis of buying the more productive method back from the operators affected for a definitive sum of money, payable within a limited period of time; set an appropriate 4M standard for the more productive method.

d. Cut the rate and let the debris of the explosion fall where it may.

The fourth (d) approach may have been more frequently used in the apparel industry in the past than the first three. It has the least to recommend it. It exacerbates irritations between labor and management. It prolongs adversary attitudes. It raises ethical questions. It causes turmoil ending either in loss of skilled personnel or in settlements for ruinously high costs and low productivity.

Earnings too high or too low mean trouble. The sooner it can be dealt with the better. Detection and report enables earlier action. This will promote harmony and not trouble in an organization, if treated with reason and with understanding of the legitimate interest of the parties involved.

Although a High and Low Performance Report will be especially valuable during the transition from manual time study to computerized 4M time standards, it will always be an important information feedback to management, and to engineering.

13

Customer Service

INTRODUCTION

The key to a continuing apparel business is prompt, efficient, and thoughtful customer service. Excellent product styling, high quality, new looks, etc., become useless if the product is not delivered to the customer complete as promised. One of the first elements of any good customer service system is to screen all incoming orders to determine if the sales promise can be kept. Quick response systems depend upon accurate linked customer service records, both with raw material suppliers and with the customers being served.

Order screening involves checking stock, raw materials, production capacity, and delivery times, as well as customer credit, prior history of the customer in respect to payment, product returns, volume of business and several other criteria.

GENERAL REQUIREMENTS

These activities fall under two general requirements.

Order Processing

Activities include the receipt and recording of orders, order changes, and subsequent reporting and monitoring of order status information. Where contracts for sequenced deliveries are made, additional records are required. Contract management activities include the receipt and recording of contracts, contract amendments and subsequent reporting and monitoring of contract status information.

Finished Goods Allocation

Activities include all efforts to allocate available product inventories in a manner that maximizes both customer satisfaction and sales. A description of typical computer customer system follows.

ORDER PROCESSING

All orders as well as other customer transactions are processed through the order processing system. This includes customer orders, all miscellaneous orders (seconds, third party orders, employee sales, CMT orders, samples, intercompany orders, closeouts, nonproduct sales, etc.), and any subsequent order adjustments. Orders are received via automated means (computer tapes, dial-up communications) and directly by phone or by mail. All orders are either keyed interactively into the system via a computer terminal in Customer Service or received directly into the computer from the customer (via tape or dial-up communications). Orders are batch processed.

Certain standard order information is optionally obtained from the product or customer files to simplify data entry. Standard values are user-controlled and defined by a business unit, customer, contract or product code level. Examples of standard product/customer information are shipping terms, partial shipment options (back order, cancel, etc.), and automatic spreading of product order quantities by color or size ranges. The system assigns a unique company order number in addition to accepting and carrying a customer order number.

Editing is performed on all orders before they are accepted. For example, edits are performed for valid customers, product codes and minimum order quantities. All lines of an order and any indicated "relaxed" orders must pass all edits to be considered accepted. Orders that fail the edits are suspended and are then accessible on-line for change and re-editing. Editing may be performed interactively, one order at a time, or in a batch mode. The order pricing option must be defined at order entry time. Prices may be obtained as standard, via reference to a product file or a contract file. Pricing may be dependent upon the status of open contracts at shipment date. In this case, no price is carried on the order when entered but the dollar value of the order is determined using the oldest open contract related to that customer and product for purposes of credit evaluation and order analysis. If the invoice price differs from the contract price, the difference is treated as an additional liability/receivable. Any specific overrides to standard prices that are entered by customer service personnel are reported for management review.

Every order entered has a credit check performed on the customer's credit exposure limit. If a customer has exceeded the credit line, the order is accepted but is reported for review by the accounts receivable department personnel. Accounts receivable personnel either place the order on "credit hold" or ignore the warning. If not on "credit hold," the order is again credit checked during the order allocation process, as described below.

Under user control, by customer option, order acknowledgements are printed and mailed, or transmitted directly to customers. The acknowledgement confirms the products, quantities and dates on the order.

An order change capability is available to customer service personnel. Under security control, all fields on an order can be changed except the order number and customer number. Customer service has the ability to change an order up until the time a picking ticket is printed. If changes are made that affect allocated quantities, that portion of the order is reallocated.

ORDER ENTRY REPORTS AND ACTIVITY

Most apparel order processing systems offered by software firms contain the following features:

a. Order Entry - Allows order information to be entered by style, color and size. Handles both single and distribution orders. Data is edited against master files before it is accepted. Credit limits are checked to determine if the order exceeds credit limits.

b. Order Correction - Allows previously entered orders to be corrected, changed or updated with new information.

c. Order Inquiry - Allows order information to be displayed showing the status of each order in whatever detail is desired.

d. Order Cancellation - Allows entire orders or specified line items to be canceled. Orders for specified styles can also be canceled.

e. Order Register - Shows information on orders entered for any specified date range.

f. Order Cancellation Register - Shows orders canceled for any specified date range.

g. Hold for Credit Report - Shows detailed information on orders being held for credit approval.

h. Hold for Confirmation Report - Shows detailed information on orders being held for confirmation.

i. Open Order Report by Customer - Shows open orders in customer sequence with quantities and dollar value subtotaled for each customer.

j. Open Order by Style Report - Shows open orders for a given style or group of styles. Provides option to show open inventory position for each style as an aid to allocation.

k. Customer Bookings Summary - Shows season-to-date and year-to-date bookings for each customer for current and prior year.

l. Invoice Creation from Open Orders - All or any part of an order can be invoiced directly without going through allocation or creating picking tickets.

m. Price Override Report - Shows the units and dollars shipped for each style at each selling price. It also shows the variance and percent variance from the standard price.

n. Projected Gross Profit - Shows the projected gross profit that will be realized from booked orders.

CUSTOMER ORDER CODING

Codes established for controlling selection of customer orders sometimes carry implications which do not always appear on the surface. Many firms develop a coding system to best serve corporate interests. This section deals with some aspects of systems and their implications.

Delivery Codes

As an example, within the clothing industry delivery codes range from no specific promise of anything at any time to a customer, to establishing a code identifying specific weeks. There are all sorts of modifications in between. The more specific the time period coded, the more detailed the preseason plan has to be and the more accurate the sales forecast and buying plan have to be. Without accurate planning and adequate piece goods delivery, detailed week by week planning seldom works. Under a Quick Response system, detailed planning is an even greater necessity and must extend into the textile mill itself.

It is the nature of the clothing industry to be squeezed between the ability of textile producers to deliver popular styles and the retailer's desire to be first with these styles. The custom in buying piece goods is to accept delivery for a month period; that is, August or July-August. As long as goods are shipped on the last day of the month, the vendor has met his schedule. Given a month's leeway on piece goods delivery, it is rather difficult to specify a weekly or even a bi-weekly manufacturing schedule for such goods unless a month's time is added to the vendor promise. Adding time to the vendor promise tends to increase inventory levels and, even then, this is often not adequate protection against slipped schedules, bad goods or shorted deliveries. Currently, apparel firms, retailers and textile mills are linking computers in an effort to overcome these inventory lags. Inventory for cutting is being replaced by just-in-time deliveries of raw materials (piece goods and findings).

Weekly Coding: Coding consists of numbering the weeks of the year. As an example, the week of June 2-8, 1986, is the 23rd week of 1986. Weekly coding has the advantage of establishing a shipping plan in advance which, in turn, can be worked back to a production plan. This method of coding permits wide latitude in the earlier portion of the cutting season, but becomes more and more of "a straight jacket" in the later part of the season. As mentioned earlier, coding works only when piece goods hit the forecast schedule, otherwise adjustments are continuous.

From a production standpoint, weekly coding tends to reduce the amount of goods which may be considered at any one time. When sizes and models are limited (as well as variations), this does not necessarily reduce height; but where customer orders must be cut complete over a wide range of lots and models, then weekly coding will definitely reduce heights and increase costs.

Ten-Day Coding: Some firms split the month into three periods of ten days each. The month is identified by one number and the time period by numbers 1 through 3. Customer promises are made for shipment by the 10th, the 20th or the 1st. When weekends are removed from this system, productive days range from 6 to 8. This means a variable shipping capacity must be calculated to determine how much can be promised within any given period. Outside of this additional calculation, the system does not appear to offer any great improvement, either in delivery promise or in improving production heights over the weekly promise system.

Bi-Weekly Coding: Coding puts two weeks in the promise period. Sometimes this coding represents a period 1-15 or 16-30. Generally, the month of shipment is represented by one digit, while the half of the month is represented by a 1 or 2. From a production standpoint, a two week period will tend to improve heights on odd sizes and less popular variations. The system provides more latitude for piece goods slippage and tends to show a better customer delivery record.

Monthly Coding: Monthly coding is closely related to the realities of piece goods deliveries. Firms which use this system, in effect, tell their customers that since they cannot establish specific piece goods delivery dates, they, in turn, cannot promise specific shipping dates within the month, except in cases where the piece goods could have been delivered for an earlier cutting period. Generally, a monthly delivery period system is tempered by a specific week promise within the month for special events, such as a store opening. In these cases, management usually makes a careful review of goods available early and reserves them so that the promises may be kept.

Monthly coding doubles the opportunity to improve height on odd sizes and variations over bi-weekly coding. Again, it should be noted that the restrictions on scanning orders for cutting are greater as the season progresses unless goods are cut ignoring promise periods.

Season Coding: Some firms use a season coding system, or ship complete by "such and such" date. This system provides the widest latitude in combining orders for cutting. Given adequate piece goods deliveries, it is definitely a production oriented system rather than a customer service oriented system. The system tends to recognize certain favored customers early while slipping less favored customers badly, sometimes even shipping fall goods on the last day of the season. Generally speaking, this system can be used only by a firm which sells goods of such a premium that the customer will take them any time, no matter how badly he is treated. For a firm which is growing and trying to develop new customers, the system would seem to present too many long term disadvantages.

General: The more specific the delivery promise made, the more risk that the promise will be broken. The smaller the time period in which goods are specified for delivery, the greater the cutting cost. Delivery coding represents management's decision on how much in production costs it is willing to pay for its effort to improve customer service. There is no doubt that if a firm can specify shipping dates reasonably closely, customer relations can be improved through promotions, repeat sales to fill in and in other ways. In the next chapter the impact of unit production systems on customer service and coding will be covered.

Static Priority Codes

Priority codes are used to call corporate personnel's attention to customer status, in one way or another. Such codes range from ranking customers by size through analyses of customer worth. Priority codes may be used not only to determine the form which customer service may take, but also as a sales management tool to evaluate coverage within a sales area and salesman's performance in respect to company profits.

In computerized cut-planning, the priority code is used as a decision device to select those customers which will not be served in the event of piece goods shortages. For this purpose, the finer the distinction drawn between customers, the better the service will be to the firm's profitable customers.

As an example, the volume of many clothing firms is often a ratio whereby under 5% of the accounts buy over 70% of the firm's production. In such a situation, unless major accounts were ranked, 95% of the customers served could be subjected to short shipments. Size of an account alone does not necessarily insure its profitability to the company, or even its desirability. Ranking customers by volume might insure two classes, those who receive service and those who do not.

On the following pages are listed 14 areas which are considered by some clothing firms in ranking customer's priority at one place or another. The weights given for the various items are subject to change, according to differing management. Not all these factors are generally in use. However, the more that are used, the more clearly defined will be the customer's characteristics. If all of them are in use, or even just a major portion, then sales management has a representation evaluation tool. This tool can be used to decide between keeping an account or dropping it. When accounts are rated annually, lists of accounts to be replaced because of unprofitability can be provided to the salesman. Some salesmen have a tendency to take the poor credit risk or poor profit account. This pattern shows up under this type analysis.

Where the foregoing group of standards have been used to rate individual stores in a chain, unprofitable experience can be called to the chain management for correction.

The way the rating system works is that each outlet is rated on each point. The results (chains or individual stores) are totaled for a composite value. Groups are then established in five unit increments such as 25-29; 30-34; 35-39; etc. Customers are tallied beneath each group to determine how many are in the group and what volume of business the group represents. All groups are totaled by both standards. Management then decides how much volume it wants in each segment and how many customers in each segment. Management also determines how many segments it wants to deal with. For example, the groups are then broadened to span 25-39 and given a ranking of B. All customers who fall in the 25-29 bracket then become B customers. From time to time, the bracketed areas may change as customers move in and out of them according to their performance. It becomes a salesman's job to move his customer's rankings up--everyone is bound to benefit.

EXAMPLE 13.1

TENTATIVE BASIS FOR RATING AND RANKING CUSTOMERS TO ESTABLISH PRIORITY CATEGORIES

Annual Volume	Rating
$ 50,000 up	1
45 - 49,999	2
35 - 44,999	3
20 - 34,999	4
10 - 19,999	5
5 - 9,999	6
2,500 - 4,999	7
1,000 - 2,499	8
Under $1,000	9

Trend	
More than last year	1
No changes	2
Less than last year	3
Abrupt drop in sales	5

Potential

Firm's Potential = Est. Suit Sales of Store ÷ Firm's Sales

(Adjust to retail/wholesale balance)

% of Sales	Rating
75 - 85%	1
65 - 74	2
50 - 64	3
86 - 100	4
40 - 49	5
30 - 39	6
25 - 29	7
15 - 24	8
1 - 14	9

Store Potential

1985 Sales divided by 1983 Sales = % Growth

Growing	1
Static	2
Dropping	9

Area Potential

1980 Population divided by 1970 Population

Growing	1
Static	5
Dropping	9

Length of Service

Year Old	Rating
1 and 20 up	1
17 - 19	2
14 - 16	3
10 - 13	4
7 - 9	5
6	6
5	7
4	8
2 - 3	9

Number of Seasons Missed

0	1	6	7
1	2	7	8
2	3	8 up	9
4	5		
5	6		

Credit Standing	Rating
AAA	1
AA	2
A Group	3
B Group	12
C Group	15
No Rating	30

Pay History	
Discounts Bills	1
Prompt Pay	2
Average 1 - 15 days late	3
16 - 30 days late	4
31 - 45 days late	5
46 - 60 days late	6
61 - 75 days late	7
76 - 90 days late	8
Over 3 months late	9

Returns in Dollar Valuation

$ Returned	$ Returned	Rating
$ 0 - $ 25	1 - 3	1
26 - 50	4 - 5	2
51 - 75		3
76 - 100		4
101 - 150		5
151 - 200		6
201 - 400		7
401 - 600		8
Over 600	6 -10	9
	11 up	

Cancellations in $

Annual $ Canceled	% Canceled	Rating
$ 0 - $ 25	1 - 3	1
26 - 50		2
51 - 75		3
65 - 100		4

101 - 150	4 - 6	5
151 - 200		6
201 - 400		7
401 - 600		8
Over $600	7 -10	9

Prestige

Well Known Name	1
Exclusively Men's Store	
Only one in County	2
One in Town	3
One of Two	4
One of Three	5
One of Four and Up	6
Mixed Department Store - Clothing	
Largest in County	7
Largest in Town	8
General Store	9

Geographical Location

Southern California and Southern Seaboard Tier	1
Second Tier	2
Third Tier	4
Canadian Tier	6
Canada	8

Class of Goods Bought	Rating
% #1 Price to Balance of Last Season's Order	
90 - 100	1
80 - 89	2
70 - 79	3
60 - 69	4
50 - 59	5
40 - 49	6
30 - 39	7
20 - 29	8
0 - 19	9

Dynamic Priority Codes

Many firms use a system to offset even less profitable customers being overlooked, by calculating the percent cut for each customer. Levels of cutting activity are established by priority group. As an example, it could be determined that in the first shipping months, 50% of priority A would be shipped; 45% of priority B, etc. The orders are then reviewed as each cutting pass is made to determine if that amount has been cut. If not, other priorities are overridden or goods available, but scheduled for later delivery, are cut.

Another method sometimes used is to establish a flat percentage cut for all accounts within a cutting period; that is, 5% the first two weeks, 10% the next two weeks, etc. This system is less responsive to customer value and may tend to drop heights under some conditions, depending upon how it is applied.

Other firms may not take a "percent done" evaluation until the eighth or ninth week of cutting, or may rely upon visual scanning of customer orders to determine whether accounts are being overlooked. Delaying review often leads to cutting rushes at substantially more than normal production costs. Regardless of what system is used, dynamic priority codes are used to insure customer service promises are kept. A dynamic code overrides both delivery and static priority codes.

FINISHED GOODS ALLOCATION

In the foregoing section, the impact of coding on customer service was reviewed;

however, most apparel firms build to inventory and allocate finished or in process goods to customers. This allocation process may or may not code customers. Priorities may be set by the rank of the salesman or by the vice president of sales. A typical industry approach follows.

The customer service department is responsible for the allocation of finished goods based upon company defined rules. Flexibility is essential to allow for de-allocation of an order up until the time it is shipped. A "forced" allocation option can be used to "push" certain orders. At the company's option, coordinate shipments can be allocated together and available inventory can be allocated on a percentage basis to each store or customer. The system optionally determines from which warehouse or distribution center the order is to be shipped. WIP inventory optionally can be considered for allocation purposes. All orders are subjected to final credit evaluation prior to allocation.

The output of allocation is the identification of orders for which picking tickets can be printed. Picking tickets are printed in the warehouse at the request of warehouse personnel. The file of picking tickets is optionally matched to detail bin/rack location records maintained by warehouse location systems prior to printing to obtain bin/rack location codes for the tickets. Other options are the ability to summarize picking tickets by store, to print multiple page picking tickets for the same order (which are used as content labels) and to print the related bills of lading. Picking tickets are numbered with the unique company order number followed by an appropriate suffix to identify multiple shipments against the same order (for example, backorders).

Information regarding anticipated volumes of orders and product quantities by ship day is available on an inquiry basis to allow warehouse personnel to plan shipments and yet delay printing picking tickets until the last moment. This practice provides additional order allocation flexibility to customer service personnel.

Picking ticket status reporting (i.e., an aged picking ticket report and an open/missing picket ticket report) is also provided.

ALLOCATION REPORTS AND ACTIVITY

Most systems firms provide the following types of reports to their apparel customers:

a. Automatic Allocation - Open orders for all styles or any selected styles are automatically allocated against inventory. The allocation rules are determined by the client and can vary with each run.

b. Allocation Report - Shows the result of the automatic allocation process so that it can be reviewed and changed if necessary. the report can be rerun as often as necessary by style or by order.

c. Allocation Adjustment - Allows changes to be made to the automatic allocation. If too many changes are required, the automatic allocation can be rerun using different criteria.

d. Reserve Inventory - Inventory can be reserved for specified orders without having to physically move goods into a reserve area.

e. Allocated Picking Slip Print - Prints picking slips for those styles, colors and sizes indicated for shipment on the allocation report.

f. Unallocated Picking Slip Print - Prints picking slips for specified orders and styles without going through the allocation procedures.

g. Picking Slip Register - Shows detailed information on each picking slip printed for any specified date range.

h. Open Picking Slip Report - Lists picking slips that have been created but not yet shipped.

i. Picking Slip Confirmation - Allows invoices to be generated for those picking tickets that have been picked.

j. Pre-billing - Pre-bills can be issued in place of picking tickets for companies that use this procedure.

SALES ACCOUNTING

The Sales Accounting functions are Invoicing, Accounts Receivable and Credit Management, Cash Collection, and Accounting Interfaces as outlined below:

a. Invoicing activities include the preparation of invoices and the related determination of prices, discounts and allowances, as well as invoice-related reporting.

b. Accounts Receivable and Credit Management activities include customer communications via invoices, statements and dunning letters, and the monitoring of receivable balances and customer credit limit to maximize collectability.

c. Cash Application activities include the receipt and posting of payments and the resolution of adjustments and deductions.

d. Sales Accounting Interface activities include the preparation of the summary level entries to properly reflect sales, cost of sales, miscellaneous adjustments and cash collections in the General Ledger for control and financial statement purposes.

INVOICING

Invoices are prepared daily by the system and, at the company's option, are printed at the division offices for mailing by billing personnel or at the home office, or are transmitted directly to the customer. Two different types of invoices are prepared, detail and summary. Detail invoices contain summary information at a ship-to level summary. Invoices are summarized by product. Computer systems have the flexibilty to prepare both and other invoice types based upon parameters specified by the company.

Normally, invoices may not be voided after issue; debit or credit memos/adjustments must be used to correct errors. Discount and sales tax calculations are performed during invoicing based upon user-defined parameters associated with specific contracts, orders, products or customers.

Debit and credit memos/adjustments are numbered using a different unique scheme and a reference to a related invoice is carried when appropriate.

The applicable invoices and an invoice summary register is forwarded to the accounts receivable department where controls are established for balancing, mailing, and filing. Invoice detail information is available, in a format similar to the invoice, via on-line inquiry throughout the month. Microfiche invoice detailed is prepared monthly.

Invoice and adjustment transactions are subsequently used by the Accounts Receivable, General Ledger Interface and Sales Analysis subsystems. Detail, summary, or bulk bill information is passed to the accounts receivable subledger, based on customer parameters, and order/contract status information is updated at the time of invoicing.

ACCOUNTS RECEIVABLE AND CREDIT MANAGEMENT

The Accounts Receivable and Credit Management function use order entry information received from the Customer Service department, from the invoicing cycle, and from credit coordination personnel in the home office Treasury office.

Reports are produced daily for the accounts receivable department providing pertinent information regarding invoices, adjustments and returns at a transaction level. This information will be used to balance the accounts receivable subledger on a daily basis. Standard monthly reports (aged trial balances, etc.) are available via hardcopy or microfiche. The accounts receivable department is responsible for mailing, optional statements and optional dunning letters to customers.

The ability to inquire into the open accounts receivable subledger is provided to accounts receivable personnel at the detail invoice level, even in the case of summary invoicing, to facilitate the evaluation of invoice deductions.

Historical information is retained at a customer level for use in making credit management decisions. Dunn and Bradstreet or other credit management information is provided by credit coordination personnel in the Home Office to assist credit management personnel. A flexible credit hold function is provided at the customer level. This, along with a flexible order credit hold function, allows division credit management personnel the ability to control orders being processed for all customers, particularly those with exceeded credit limits or those for which prepayment before shipment is required. Periodic consolidation of credit exposures at a company-wide level are requested and reviewed by credit coordination personnel in the home office to monitor total exposure versus a company-wide limit for large, common customers.

CASH APPLICATION

Most cash collection is currently accomplished through the use of bank lock boxes. The system has the flexibility to allow the company to specify the lock box to be utilized by bill-to, customer, state, or geographic area. The collection banks forward deposit totals and copies of remittance advices and checks to the company offices daily. As bank deposits are received, each deposit is balanced to the included check copies before any cash is applied to individual accounts.

Cash application is an on-line function. Payments are applied to either detail, summary, or bulk bill invoices as appropriate for that customer. The system provides for partial payments and also allows direct writeoffs of small amounts. Limits are established for writeoffs and "audit" lists of writeoffs are prepared for management review to control the writeoff feature. Payments of multiple invoices with the same check can be applied on a specific item basis, on a FIFO basis or can be posted only "on account" for subsequent resolution. The FIFO option excludes invoices coded as "in dispute." A number of collection status codes such as "in dispute" are provided in the system.

Payments from unidentified customers are posted to a special account for subsequent resolution. Each deposit batch must balance upon completion of posting.

e. Gross Profit Report by Customer - The gross profit on net units shipped (sales less returns) is computed for each customer for season-to-date.

f. Gross Profit by Style - The gross profit on net units shipped (sales less returns) is computed for each style for season-to-date.

CONCLUSION

Modern computer systems interface with the bulk of all corporate activities. Customer service systems require data on customers, inventories, various sales analyses and delivery status. The system feeds the financial data systems and influences corporate planning and merchandising policy.

14

Summary—Computer Controls

GENERAL

This chapter contains a brief topical summary of a typical integrated system as currently offered by several software firms. The summary outline is then followed by a discussion of the impact of "just-in-time" deliveries and "quick response" customer service on existing systems. Basically, computer systems must undergo continuous change in order to provide appropriate information to corporate management. As business conditions change, so must systems. It has been said that good systems engineers spend most of their time obsoleting their own work. Whether or not this is true, there is no doubt that the textile-apparel complex is changing the way it does business. This means new computer control systems.

INVENTORY MANAGEMENT AND ORDER ENTRY

Inventory Management

Inventory management systems are designed to obtain concise and accurate information for control and planning of planned goods, issues, cuts, projections, work in process and finished goods. These systems provide summary and detail analysis for both production and accounting control systems.

System Highlights:

- Multiple warehouse locations for in transit and consignment
- Finished and third quality inventory
- User defined size and color description
- Style analysis by company and location
- Manufacturing time phases with component levels of cuts or finished goods
- On line display of all inventory transaction history
- On line display of inventory by location
- On line display of order allocation detail by style and color
- Book vs. physical reports in a complete inventory sub-system

- Missing tag lists with automatic reconciliation of book to physical inventory
- Complete cut and sold reports
- Inventory valuation of actual, standard, or frozen standard cost
- Two step cycle accounting with purchase price variance analysis
- Detail and summary inventory turnover and usage analysis
- Uncosted purchase order analysis for receiving goods
- Three way match of purchase order, receipt and invoice
- Real time interface with production, purchasing and order entry
- Supports multiple issues and partial picks from the manufacturing interface
- Interfaces with accounts payable.

Customer Order Entry

Customer order entry systems are designed to capture orders in an interactive real time environment. These systems normally provide built-in credit checking and inventory techniques. Systems are designed to increase shipments, reduce backorders, and provide production and accounting with timely and efficient projections.

System Highlights:

- On line order entry and maintenance
- Optional back ordering and inventory checking by warehouse and location
- Allocation of coordinates, both complete and partial
- On line validation by style, size
- Order quantity crossfootings, and summary totals are supported by line
- Flexible deallocation and adjustment of line items
- Warehouse pick tickets can be generated at random
- Style substitution and modifications affected prior to pick slip generation
- Pick slips can be generated by warehouse and location
- Bulk picking reports and manifests.

Billing

System Highlights:

- On line invoice release via order, pick slip or direct billing
- Data maintainable at release time
- User controlled back orders and partial shipments

- Post shipment, returns entry and update

- Audit trails to A/R

- User defined inquiries and bar graphs

- Automatic posting to the general ledger.

Sales Analysis

System Highlights:

- Invoice register with sales, costs, taxes, freight, special charges and comments

- Gross profit by customer, style, salesman

- Detail analysis by customer showing units, sales, and costs by month and year to date

- Sales by location, year to date, and prior year to date

- Salesman's commission, dollars year to date and prior year to date

- Customer sales dollars, year to date, current, and prior year to date

- Individual style shipment transactions for shipping performance and invoice detail analysis

- User defined reports and bar graphs.

FINANCIAL SYSTEM

Financial modules are designed to allow any business organization to take advantage of a flexible and interactive multi-company, multi-division system.

All modules usually interface directly with inventory, purchasing, billing and sales analysis, with multi-country currency capabilities. Interface via personal computers or main frame provides for the uploading and downloading of data for spreadsheet analysis.

General Ledger

System Highlights:

- Account numbers are user defined
- Separate companies can have different closing periods
- Consolidating of profit centers
- Adjustments can be made to prior periods and years
- User defined financial reporting via a financial retrieval system
- Year to date and previous year budget comparisons
- Recurring journal entries and automatic entry from all sub-applications
- Will process data for a calendar week, year or defined accounting periods

- Support prior year and prior period adjustments
- Journal entry inquiry for current periods and prior year
- Automatic distribution of overhead expense to user departments
- Automatic generation of accruals and reversals
- Historical budget data available.

Accounts Receivable

An A/R system is designed to facilitate the control of subsidiary detail. It enables the accounts receivable department to speed up credit and collection, recognize delinquent accounts, and provide a clear and concise cash analysis by company division and/or profit center.

System Highlights:

- On line posting of debits and credits
- Multi-company and multi-cost center support
- Line item or mass posting of cash
- Multiple invoice terms with editing of discounts taken
- Variable aging including future with open item or balance forward capabilities

- Charge back control to general ledger
- Aging by customer, company, salesman location or consolidated aging
- Complete audit trails for all transactions
- On line inquiry by customer with complete credit history
- Supports multiple terms and variable invoice due dates
- Credit analysis
- Automatic posting to general ledger.

MANUFACTURING

The manufacturing system is designed to provide for the control and tracking of items that are either, made to order, or for inventory:

a. Planned style cut--cut planned not issued to cutting room

b. Special process--component processing, dyeing, knitting, etc.

c. Issued style cut--issued to cutting room not reported cut

d. Cut style--actually cut adjusted for yield in work in process

Manufacturing applications are integrated with inventory management, purchasing and order processing.

Shop Floor Control

This allows for detail planning, scheduling, and highlighting of occurring problems and gives management control over the production, material scheduling and capacity planning.

System Highlights:

- On line modification of components for substituting routines for an order
- On line labor posting with shortage lists by component and order
- Detail inquiry of open shop orders by item or order
- Prints picking lists, shop packets, and labor tickets
- Supports setup and running of machine hours
- Supports work center loads by combination or separate setup of machine hours
- Efficiency report by work center and employee
- Close operations automatically
- Provides a separate shop calendar and separate outside operations
- Backward schedules operations based on shop order, due date, and schedule date for material issues from stock

- Provides an order shortage list by component and handles alternative routings and alternate descriptions

- Job and product costing

- The cost accounting system module pinpoints manufacturing and purchasing for efficiency as well as variances for accounting and production control. It gives management necessary information for pricing and production of its products.

Cost Accounting

System Highlights:

- Cost analysis by item, employee, department, and work center

- Provides labor and overhead costs

- Standard, frozen standard, and actual cost of material and labor

- Automatic allocation of overhead by hours or pieces

- Variable costing of overhead by work center

- Automatic loading of standards with last year's costs

- Average or last order costing and work in process costs

- Cost simulation and material usage costs

- Gross profit reports.

MRP/MPS

Material requirements planning and master production scheduling give top management, purchasing, and the production department the information necessary to control their every day business operations. It evaluates detail production and purchasing and ties the detail operations and business planning through the use of master scheduling.

System Highlights:

- Bucketless MRP with regeneration capabilities
- Supports firm planned orders
- MPS planning reports with availability
- MPS simulation and variable planning horizon
- Time summarization of MRP and MPS data
- Real time MRP planning with inquiry facilities
- Summary production with revenue and expense projections
- Suggested rescheduling of purchase orders, shop orders, and firm planned orders
- Supports forecasting and consumption with customer orders

- Real time planned and pegged order inquiry

- Capable of lot sizing and order policies

- Action messages with dates

- Provides for the multiple release of planned and firm planned orders to the shop floor.

Bill of Materials System

The bill of materials allows for the retrieval of product information used in a wide variety of costing needs by production and cost accounting. It allows for immediate on line posting and access to bill of materials data.

System Highlights:

- On line duplication and modification of bills

- On line posting of all maintenance for additions, updates or deletions

- Flexible decimal places available for quantity

- Item inquiry and multi-level where used inquiry

- Bill of material for purchased items

- The mass replacement of components and structures

- Multiple component for the same item on a single bill

- Indented and single bill of material available in inquiry or in report form

- Calculates cumulative lead times.

Capacity Planning

Capacity planning will pinpoint loading problems at critical work centers. It allows the user to improve routing, scheduling, and will identify backlogs and bottlenecks.

- Allows for the setup or run of machine hours by work center

- Provides detail capacity planning requirements on line or in report form

- All orders are passed with requirements for actual or planned

- Work center loads are summarized in a flexible time period.

- Rough cut planning and pegging provided from the master production schedule

- Provides a backward scheduling of planned operations

- Generates a bar chart display of capacity requirements by time period.

INTEGRATED BUSINESS SYSTEM SUMMARY

Most software houses purport to sell and install an integrated business system.

In general terms, they are correct. However, each firm has operating, sales, financial and other conditions which require adjustment or "tailoring" the offered package. Sometimes this tailoring can be quite extensive and very expensive. Totally integrated business systems combine critical operating and financial systems of manufacturing accounting and distribution. This integration permits the apparel manufacturer and importer to control the dynamics of their business in a new and sophisticated way. Current software solutions are comprehensive, interactive and end user oriented. They provide the user with a business perspective which could only be obtained by combining several stand alone systems.

IMPACT OF SINGLE UNIT PROCESSING ON SYSTEMS

Introduction

According to the February 1985 issue of *Bobbin*, fifty of the country's leading retailers endorse quick response, i.e., shortening the time cycle between order and shipment. Buyers and merchandisers from major retail firms agreed that within the next three years retailers will be giving domestic manufacturers less time to deliver initial shipment and reorders than they are at present. Two out of three firms interviewed agreed they would be willing to pay a premium for the faster delivery.

Response time reductions of two weeks equated to a 15% increase in retail sales in this survey. Retailers generally agree that shortened "stock out" time through quick response reorders would increase business, lengthen the selling season and reduce markdowns. A 10 week delivery cycle from order to shipment receipt is considered obsolete for fashion wear goods. Some domestic firms are actively shortening their delivery cycle time (2-3 weeks over past three years) and are moving toward even better service.

What does it take to stay in the $10 billion retail sales market represented by the fifty leading retailers (2200 outlets) interviewed in this survey? It requires a new order of:

a. Manufacturing, equipment and systems

b. Planning production

c. Ordering raw materials.

Anything less will not be enough.

Single unit conveyorized processing systems can be linked to any type of spreading and cutting system--ranging from hand cutting through computer controlled cutting by knife or laser. Many of these systems have the option of tracking cuts and of generating payroll data.

High ply cutting is normally associated with a mobile dolly transfer system. Large cuts are often associated with building large finished goods inventories in anticipation of customer orders. A mobile dolly system usually causes an apparel manufacturer to increase the inventory investment in raw materials, in cutting, in stocks on the plant floor and in the finished goods warehouse.

Many firms associate low ply cutting with cutting to customer order. This is most noticeable in the tailored clothing industry. Lowered heights increase cutting costs but may reduce inventory risks. Low ply cutting associated with a mobile dolly system can increase bundle handling and other production costs without necessarily speeding customer deliveries.

Single unit "one-at-a-time" processing, properly implemented, offers the apparel manufacturer the broadest opportunity to provide quick response to their retail customers. In order to implement a single unit system, new equipment and systems must be acquired, tested and proven. Closer bonds with textile and trim producers must be forged. Exact, correct information must be provided to salesmen, vendors and customers.

Manufacturing--Equipment and Systems

Single Unit Processing - Background Example: A typical five pocket jean usually is completed in thirty-five (35) to thirty-eight (38) sewing operations. Normally the standard allowed hours range from eighteen (18 SAH/100) per one hundred units to twenty-two (22 SAH/100) depending upon the fabric, the make and plant practice. Expressed another way, a typical jean requires from ten plus (10.8 SAM) to thirteen plus (13.2 SAM) standard allowed minutes of construction time to make.

Single unit processing requires a balanced continuous flow of work through each operation. The flow may not necessarily require processing a single operation at each station, but could include several operations utilizing the same equipment and sewing skills. Given a sewing production rate of twenty (20) seconds per unit [3 per minute, 180 per hour, 1440 per 8 hour workday], then the work stations within a single unit system must be balanced to deliver three units per minute. Assuming a balanced line and a one minute interval of process time between stations, then a completed garment will require between forty-six (46) to fifty-one (51) minutes of process time. Tightening the in-process standard by fifty percent results in a process time ranging from twenty-nine (29) to thirty-two (32) minutes.

In checking users, some apparel firms reporting nineteen (19 SAH/100) standard allowed hours per one hundred units reported throughput times of under thirty minutes. Other firms reported "one or two day" in process time. A three units per minute pace applied to the above jean example establishes potential throughput times which range from twenty-three through twenty-six minutes per unit. Single unit processing (including cutting) can be viewed in another way. Given a production line geared and balanced to produce three garments per minute, the sewing (add 1-1/2 minutes cutting time to the sewing cycle time to obtain complete cycle) of a 231 unit order for a customer theoretically can be completed as follows:

Example: 231 ÷ 3 = 77 + 23 (or 26)
= 100 mins (to 103) or 1 hour 40-43 mins.

In theory, a line balanced at 2880 units per eight hour shift could complete the same order as follows:

231 ÷ 3 = 77 + (35 ÷ 6 + 11 SAM(16.6)) = 93.6 minutes (alt. 96.5 min) or 1 hour 34-37 minutes.

This means that 12.6 store orders of this size could be complete per 8 hour day, once the pipe line is filled (2880 ÷ 231 units). These figures assume:

a. All piece goods and trim on hand

b. Proper line balance including

    (1) reserve equipment (roll in and out)

    (2) utility operators

    (3) properly computed delay allowances

c. Ability to pack by customer order

d. Ability to cut by customer order

e. In line quality control

f. Thread free garments

g. Close supervision

    (1) monitored production

    (2) adjusted line balance.

Realistically, a jeans shop should try to gradually reduce its in-process time from the industry ten to twenty day cycle to a one to two day cycle.

Single or Low Unit Cutting

Cutter Background: Lasers have been used in apparel cutting (one high) for several years. Recently several equipment firms have developed new laser cutters. Some of these are now entering the market. Prices on these cutters range from $45,000 to $150,000, depending upon the brand, the laser power and the support system. These systems, properly used, have demonstrable piece goods payback. This report will include some figures from a jeans example to illustrate payback potential.

Lasers cut denim and denim blends accurately and cleanly. Since work is clean, folder and seam allowances can be reduced. In addition, the laser cutter can cut exactly on common seams, does not require space for notching and can provide precise drill holes. The laser is a single or low unit cutting system. It is capable of handling small lots, reorders and private label cuts as well as special sizes, tailored garment cutting and a wide range of "exception" activity. A laser system has been tested at 3 ply high cutting using a 600 W laser. A 1000 W laser may be used to improve cutting speed.

Tests were conducted on a conveyor belt system laser cutter. This system cuts material up to 72" wide on a sectioned stainless steel belt. It operates at about 1-1/4 yards per minute or the equivalent of about one adult jean a minute (± 480 per work day). Lightweight cotton goods were test cut 8 ply high. Tests on denim have been made up to 6 ply with 3 ply on the 600 W laser producing the best results.

Marker/Grader

Of particular interest are the new designer/grader/marker making systems which provide floppy disks to drive the laser cutter or to produce a marker. Lectra also permits pattern changes on the tube, cuts the prototype marker so that new test garments can be made more rapidly. At Basset Walker (Apparel Industry Magazine, March 1985), an untrained computer operator (a designer) was able to operate the system successfully within a month.

An automatic marker system packs garment piece parts into specified widths. In a trial of 100 parts (coats and pants, men's clothing--see Apparel World, June 1984, page 52), the computer made a 74% efficient marker in seven minutes. It took a top marker maker using the system fifteen minutes to make an 83% efficient marker. When the marker maker took eight minutes to improve on the seven minute marker, he produced an 84.2% efficient marker. (Note: Pieces were tilted to gain the improvement.)

Production Cutting Systems

Spreading can be eliminated if proper piece goods holders, power drives and edge guides are installed. Material handling is reduced to once a shift if 600 yard rolls of denim are specified and used. Standard rolls improve shade and width matching for release of goods to a cutting room. Cutting is eliminated by the laser except for an equipment adjuster. Shade numbering is eliminated if dye lots are kept together. Bundling will stay with different duties if the laser cutter is associated with a single unit processing system. Bundlers will have to pick up parts, match sizes and place a size in the proper conveyor unit. A marker-laser operator and a bundler will be the primary expense. Indirect cutting labor costs should be about the same with some personnel shifts. Prime advantage would be uniform, precise cutting, with good notching. Next major advantage is the immediate capacity to adjust markers. End loss can be eliminated using the large roll approach. Width loss can also be reduced or eliminated through the purchase of uniform piece goods. Marker savings can be gained from improving loose markers by using automatic marker system and adjusting the marker manually.

SINGLE OR LOW UNIT SPREADING SYSTEMS

With a conveyor laser cutting system, it is possible to eliminate the conventional spreading function. Combine this with "just in time" piece goods and trim, deliveries inventories drop substantially.

If an apparel firm buys three to five roll blocks of piece goods 600 yards long from the same dye lot and beam, width variations would be minor and shade variations would be controlled. Rolls would be loaded once a shift. Not only would this system save floor space, but end and width loss would be minimal since there are no spreads cut to length and piled high. Continuous feed and continuous cutting of single dye lots from the same beam will provide the opportunity to cut shaded matched goods for individual customer orders. This action can improve the quality of shipped goods and should improve the retailer's rack display appearance when hung together.

SINGLE UNIT GARMENT SEWING

Single garment unit conveyor systems have been available and in use for a number of years, primarily in Europe, Canada and Japan. Displays at the Bobbin Show of these systems have increased from one in 1980 to a half dozen in 1984.

The N.C. State Eton system includes not only a conveyor system but also several special purpose work stations to reduce operator positioning and handling time. By reducing the WIP level, quality is easier to control and maintain. Random quality checks are easily performed by supervisors as a part of their normal daily routine. Therefore, quality problems should not affect as many garments as a progressive bundle system. Completed work is readily visible and easily observed at each station. Errors can be caught earlier. With the bundle system, bundles are retied and placed on dollies. Bad work may be concealed since it is not readily visible.

CUT/SEW SINGLE UNIT SYSTEM SUMMARY

Most present apparel cut/sew systems are not designed to handle small lots, reorders or private label cuts. A single unit processing system, cutting three high and sewing one unit at a time provides the ability to cut/sew/pack small lots (assuming material on hand) in one to two days, depending upon order size. The example reviewed here will not only cut/sew 2,880 daily units (two shift cutting, one shift sewing) but will also provide improved markers. Analysis of cutting operations prepared by three vendors and several engineers in a series of reports starting in June 1984 and running through October of that year in several plants, indicates that there are generally four cutting room problems that can be solved or helped by using improved systems:

a. Materials utilization

b. Poor markers

c. Obsolete spreading equipment

d. Out-of-date tables (unable to handle heavy, large rolls).

Lectra equipment will help solve a marker quality problem. Lectra's automatic marker making function has been observed in use. It was tested and did work on a jeans spread. Markers can be improved by "tilting" pieces in the marker.

In addition, marker width loss can be substantially reduced by making new markers as widths change. The automatic marker making feature can produce a new marker in under fifteen minutes. Cutting 3 to 5 ply high should also tend to reduce width loss since it is easier to match widths on low cuts than on high cuts.

Tables and spreaders are eliminated by utilizing a spreading rack which will have to be designed and built. This action frees floor space for additional sewing activity. Cutting to order also frees warehouse space which can be placed into manufacturing activity.

The unit processing system offers the opportunity to cut to customer order in a very practical way. However, this system requires more rapid and accurate planning. The "just-in-time" concept offers the potential of not only lowering costs and inventories but also the freedom to add sizes, special styles, etc., without increasing costs (multiplying SKUs at no inventory risk).

PLANNING PRODUCTION

Single unit processing requires tight coordination between the marketing and manufacturing segments of an apparel firm. When single unit--single customer--production begins to incorporate "just in time" deliveries of piece goods and trim, close cooperation between all corporate elements becomes even more important. Those firms which are able to act promptly in response to customer needs will reduce inventory investments, in raw materials, in process and in finished goods.

In order to develop an organized approach to cutting to customer order, a firm must evaluate three primary areas:

a. Functional. Those areas that pertain to specific departments with specific duties.

b. Operating. Those that relate to procedural activities assigned to functions.

c. Controls. The assignment of responsibilities for activities to be carried by specified functions within proper time frames.

"Cut to order," single unit, "just in time" production offers procedural and operating opportunities to bypass all the old accumulated, encrusted operating habit patterns. Most present systems of any established firm represent patching or grafting a computer customer service system to satisfy continually changing needs. Most customer service systems are still oriented toward serving customers from a large inventory using a sequential production system. Under present sequential production systems, inventory reserves are established at multiple production and customer service levels; i.e., dollies at each work station, cut goods ahead of sewing, roll goods ahead of cutting, finished goods ahead of customer order picking, etc.

Customer service under the <u>single unit</u> production concept can embody the direct transfer of customer order data to the laser cutting system. The change in concept would require a customer planning/ production planning system similar to those in use in the tailored clothing industry. Under the proposed low unit cutting system, customer orders would be combined into a continuous eight hour three-high cut. A computer printout of how the orders are combined in the continuous three-high full-shift marker would be used to fill, label and box orders at the end of the line.

Beyond these changes, a new type of seasonal calendar and capacity planning system needs to be developed. This report will outline some of the procedural differences between normal current "batch" production systems and the proposed "cut to order," single unit, "just in time" system.

PRESEASON PLANNING/PRODUCT DEVELOPMENT

Time Phased Merchandise Plan

A master planning calendar is still required. However, it must embody at least two new features.

a. Direct liaison with textile and trim vendors to establish a seasonal "just in time" delivery schedule. (Some textile firms link their computers to apparel firms.)

b. Direct work with merchandising and sales functions to establish proposed customer delivery plans (including reorders). Sales plans should include balanced styles and stocks in promotable saleable quantities. Styles should be planned and spaced for delivery so as to maintain a "fresh" stock look on the store floor. Work with retailers should also include periodic size review by salesmen to establish fill-in reorders. (Some retailers use computer-to-computer ties between retailer and manufacturer as the method of reordering.)

Purchasing/Quality Control

Expansion of the purchasing/quality control role to include some shifting of the manufacturing/merchandising production plan. Changes would be made to accommodate trim or piece goods delivery deficiencies in planned "just in time" delivery schedules. Ordering raw materials will be substantially changed. Very close links with venders will be required.

New Product Coordinator

Apparel firms must establish a new product coordinator to insure early complete specifications for production of samples and for generating an overall store by store sales plan.

Other Activity

Previous recommendations in the areas of new product development, product costing, forecasting, quality control specifications, design and material standards control are unaffected by single unit processing. (See the 1982 and 1983 AAMA reports on Master Planning Systems.)

PRODUCTION CONTROL SYSTEMS

Introduction

Normally, merchandising bulk plans goods and styles to be cut, sewn and delivered during a specific month. Salesmen are advised that 10,000 Style X will be customer available, say 8/1-15/85. Under this practice, as orders are received, the available units are reduced until all units are sold. Usually, over-solds are advised of a later availability. This merchandise bulk is then translated into a manufacturing plan.

Basic Planning

Raw Materials: Under the present delivery time lags imposed by the textile industry upon the apparel industry, apparel producers do the early piece goods and trim style forecasting. Since the apparel firms must estimate what can be sold long before they are able to deliver their products, they assume the bulk of the inventory risk. Once a raw material commitment is made, then the apparel firm has to generate interest in the base material, by styling, alternate uses, promotions, etc., so that inventory losses are reduced or eliminated.

It is assumed that there will be little or no immediate change in present textile delivery activity. It is assumed that plants can be preloaded by computer (i.e, loading styles one season in advance on all present systems including skills and equipment at the plant level). Once the plants are loaded, the purchase order for goods instructing delivery just prior to use should be issued specifying quantities by cut week.

A receiving schedule for each plant and/or for each raw material storage area would also be issued at the same time. Where possible the apparel purchase/ receiving schedule should be linked to the textile production planning computer program. As changes occur the receiving schedule would be reissued. (Note: Some firms issue company numbered receiving tags for each style and roll ordered so that the receiver can check yards, width and shade on a portion of the receiving tag. This is then entered into the computer.)

Customers: Under present delivery requirements imposed by retailers upon the apparel industry, the confirmed, sized order is delayed as long as possible. Single unit processing will require confirmed quantities and will permit last minute sizing by retailers. Present normal time lags between initial order and initial delivery average ten weeks on fashion goods. Some delivery requirements are as low as four weeks (on girls' wear). The thrust of retailers is to shorten the domestic response time even more.

Most apparel producers, faced with a one and one-half month to three months piece goods delivery cycle and a similar length domestic manufacturing cycles, produce to forecast rather than to customer order. Producing to forecast tends to generate finished goods inventories which seldom exactly match customer orders.

Single unit processing permits cutting styles and sizes to customer order, thereby reducing the finished goods inventory risk. To accomplish this objective apparel firms will still have to have the piece goods and trim flow in place but will not have to commit production prior to the sized order receipt.

Cut Planning

Present Systems for Planning Goods for Cutting: Many firms do not preplan individual pieces of goods against planned cuts to minimize end loss. Most present systems generate a quantified manual cutting ticket listing piece goods needed. Goods are usually issued until the total yardage is covered (with some reserve for ends). Cuts may contain several sections of varying heights which create more ends.

Current cutting ticket heights are usually balanced at traditional dozens multiples such as 36, 48, 60 or 72 pairs high so as to create standard bundles for a progressive bundle handling system. These bundles may be broken into 30 or 36 unit bundles during the assembly process as bundle bulk increases beyond the capacity of a standard dolly system.

Altered Systems for Planning Goods for Cutting: Low ply cutting requires a cutting ticket which lists the sequence of 480 sizes cut over an eight hour shift. This cutting ticket can be for 1,2,3,4,5 or any other height for which a laser produces an acceptable cut at an acceptable speed (i.e., on adult jeans a size a minute). Total yardage can be computed for the eight hour shift (i.e., 480 units x 1.25 yds = 600 yds). The number of 600 yard rolls needed in this example would be pulled from the same dye lot. If the cut is three high, then three rolls of 600 yards would be pulled and loaded prior to the start of the shift.

The standard starting bundle could then be three units which can later be split to one unit as bundle bulk increases and semiautomatic positioning becomes more difficult for the conveyor system. The basic process is almost the same as the present process. Cut size is reduced. Present cuts are in the 2,500-3,000 unit range. Laser cuts (3 high) would be in the 1400-1600 unit range depending upon age group, gender, size and style selected for the cutting.

Present Cutting Ticket Data: As mentioned above, present cutting tickets specify the piece goods to be used. Trim requirements are usually implied. A copy of a cutting ticket may be provided in advance to the trim room where the volumes of those items needed are calculated by the trim clerk. Labels, tags, rivets, pocketing, etc., may be pulled, and the volume pulled may be noted on the ticket. Some items may not be provided because they are already available on the plant floor. Present cutting tickets do not usually list the private label or cut to order customer. As a result, completion of these units is delayed and is double processed (stored unfinished or unlabeled and then picked for finishing with all the time delays and costs inherent in such a system).

Altered Cutting Ticket Data: In listing the sequence of units to be cut, the low ply cutting ticket can also list the customer for which the sizes cut are intended. Special trim requirements can be accumulated so that the trim clerk can pick the material for sending to the cutting room along with the piece goods. A sample ticket may look like this:

	PIECE GOODS		Sizes ⟶			
	ROLL#		32-32	32-34	28-26	- - - - -
Ply 1	6621	Customer #	12345	12345	12345	etc.
Ply 2	6622	Customer #	12345	12345	12333	
Ply 3	6623	Customer #	12345	12345	12640	
	Time Cut 6/13/85		8:01	8:02	8:03	8:04

TRIM DATA SUMMARY LABELS		BRADS	BUTTONS
127 Cust#	12345	etc.	――――――
677 Cust#	12337		
480 Cust#	12301		

Cutting tickets can be expanded to cover pick/pack instructions and shipping labels. Both the trim summary and the pack summaries can be printed separately. Cutting tickets could cover 15 minutes, 1/2 hour, 1 hour, 1/2 day or 8 hours, one customer, two customers, or whatever seems reasonable.

SETTING UP THE NEW SYSTEM

As mentioned in the preceding section on equipment, it is estimated that $150,000 will be the cost of plant preparation, equipment installation and management training. About six months will be needed to bring a production line into smooth operation according to the experience of those who have made an installation.

ORDERING RAW MATERIALS

Line Sheets

Preparation of line sheets to establish styles, quantities and delivery dates is a basic merchandising function. Once such a sheet is completed and approved, it becomes the principal piece goods planning document. The line sheet is also used to establish trim consumption requirements. Trim requirements are derived from consumption history and from sales estimates.

Purchase Orders

Once the raw material requirements have been quantified by delivery date and by fabric, trim, and related items for a sales season, a delivery schedule can then be established to minimize raw material stocks while still satisfying the proposed delivery schedule. Under the Motor Carrier Act of 1980, it became easier to arrange for special service and rates. Not only did this act reduce shipping costs, but also made it possible to use the just-in-time inventory technique.

This technique reduces inventory purchase and storage costs by mating inventory and production equipment as needed. Additional legislation is under consideration which will complete trucking deregulation begun in 1980.

Purchase orders can now reflect quantities needed by specific day, week and month. Carriers can be designated to reflect a season's negotiation.

Conversion of Bulk Orders to Customer Order

The production plan together with the purchase schedule is reduced under present practice as customer orders are received. A gradual conversion of bulk planning to customer order response activity will occur as the unit production system is installed. Under the short cycle production system proposed, it is possible to accumulate daily cuts dedicated to specific stores and customers since units can be planned to the minute, size by size for each day.

Trim Requirements

Merchandising selects normal trim items to be purchased and recommends vendors. Special trim items, labels, billboards, buttons, rivets, etc., may be selected in agreement with a customer. These private label items should become a part of the customer shipping information.

Manufacturing normally establishes min-max quantities of trim items. As the in house stocks are reduced, reduction should trigger a computer printout of those items needing to be ordered. The printout is reviewed and reorders, if needed, are made.

As noted in a preceding section, cutting tickets can be modified to show special trim requirements as well as standard requirements. Under the unit processing system, all trim must be ready and be placed with the units as they are cut. This means a picking list for piece goods and trim must be prepared at least the day before cutting so that all elements needed for the garment will be ready as production begins.

CONCLUSION

"Just in time" deliveries and unit processing systems cutting to customer order require different computer control systems. Quick response is the immediate aim of many textile and apparel firms. Controls will tend to reach down to the detail plant floor level and may bypass much of the summary report data outlined in preceding chapters. The plant report card will become timely shipment.

In examining or establishing new systems, an analysis of the company's overall direction is necessary. Many firms have spent substantial funds on "packaged" or "modified package" systems only to find the system to be obsolete. Particular care in system selection is needed whenever an entire complex, such as fibers, textile, apparel and retailers, is changing its way of doing business.